Chemical Process
Simplification

Chemical Process Simplification

Improving Productivity and Sustainability

Girish Malhotra, PE

WILEY

A JOHN WILEY & SONS, INC., PUBLICATION

For general information on our other products and services or for technical support, please contact our Customer Care Department within the United States at (800) 762-2974, outside the United States at (317) 572-3993 or fax (317) 572-4002.

Wiley also publishes its books in a variety of electronic formats. Some content that appears in print may not be available in electronic formats. For more information about Wiley products, visit our web site at www.wiley.com.

Library of Congress Cataloging-in-Publication Data:

Malhotra, Girish, PE
    Chemical process simplification : improving productivity and sustainability / Girish Malhotra.
        p. cm.
    Includes bibliographical references and index.
    ISBN 978-0-470-48754-9 (cloth)
1. Chemical processes.   I. Title.
    TP155.7.M355 2011
    660'.28—dc22   2010031080

Printed in Singapore.

10 9 8 7 6 5 4 3 2 1

To

Indu
my known and unknown friend

Contents

Preface

An opportunity to write about one's experiences and methodology to review, develop, solve, and simplify processes can be a challenge. However, if one is taught and trained by excellent teachers, colleagues, and supervisors, the challenge can be fun.

I was presented with this opportunity in 2008 and I am thankful to the people associated with my career. My thanks are to many but a few people stand out. Among teachers are Mr. Shiv Narain Das Gandhi, Mr. J. P. Srivastava, Professor D. D. Arora, Mr. B. N. Tandon, Professors Dr. R. D. Tiwari, Mr. R. P. Singh, Dr. J. B. Lal, Mr. S. B. L. Sherry, Mr. Himmat Singh, and Dr. W. N. Gill. Among colleagues and supervisors are Dr. Chester Snell, Mr. Edward Cave, Mr. Keith J. Conklin, Dr. A. K. Nanda, Dr. F. W. Sullivan, Mr. F. E. Butler, Mr. R. D. Hardy, Mr. K. Haber, Mr. Juan Jarufe, Mr. C. G. Ivy, Dr. B. G. Bufkin, Dr. R. E. Medsker, Dr. Richard Thomas, Dr. Charles Kausch, and Mr. Steve Waisala. I am also thankful to all the operating colleagues who shared their insight and supported me. As a matter of fact, since the operating personnel are the ones who must live with any design flaws, they have the uncanny ability to simplify the operating methods; their insight gave an excellent perspective of processes and the challenges they faced. Their perspective and creativity was always valuable.

We humans are creative individuals, and due to the rigors of 24/7/365 operation, we want to have processes that are repeatable, require minimum attention, and deliver quality. In this effort, well-trained operating personnel also want to minimize pollution, effluent discharge, and maximize on-stream-time. Similar operational strategies apply for batch processes. In their effort an experienced operator is the best friend who can critique a process before it is commercialized.

There are other methods and ways to address other and similar challenges. Methods discussed here have helped me to rationalize, solve, and create solutions.

The book's title is very much what I believe for any manufacturing operation. Our creativity, imagination, beliefs, persistence, and enthusiasm are the key to simplicity and sustainability of the process. These are not the easiest things to share and sell, but are the key to one's success. Increasing "profitability through simplicity" has been my mission and it is an exhilarating experience. At one point in my working career, during a discussion I realized that we figuratively work for a corporation through its hierarchy, but we really work for ourselves, as we have to deliver our best. Results of our efforts might not always be to our expectations, but the experience provides a valuable lesson that no university or school can teach us that. No one can take these experiences from us.

The experiences I have had at each of my employers and with my clients have been unique and cannot be duplicated and/or repeated. After discussions with some of my colleagues, we agree that a "can do attitude," camaraderie, and creativity rubbed off on each other and enabled us to do things that previously would have been considered impossible. I have to thank all of my colleagues who helped me in my learning experiences.

I also thank my parents, Mr. and Mrs. G. D. Malhotra, who gave me the freedom to do what I wanted to do, considering the culture and environment of growing up in India. I would also like to thank my brother and his wife, Mr. and Mrs. Umesh Malhotra, for their unending support and encouragement. Last but not least, I am thankful to my wife, Dr. Indu Malhotra, for her objectivity, simplicity, and cheering that helped in this whole process. Our children, Mrs. Malvika Paddock and Dr. Rohit Malhotra, in their own ways helped me to stay on track. I am delighted and thankful.

Thanks and cheers,

Girish Malhotra

Chapter 1

Process Simplification: Basic Guidelines

In today's competitive world it is necessary to have the highest-quality product at the lowest cost. In addition, products must be safe and environmentally friendly. Most manufacturing processes, even the best, can be improved to reduce cost and enhance product quality. At times process simplification/improvement may seem like a formidable task, but every step in that direction is a satisfying experience. Since we are trained to apply our knowledge and experience to achieve such objectives, these tasks should be easy.

Process simplification invariably results in the following:

- Improved product quality
- Reduced waste
- Reduced batch-process cycle time
- Reduced raw-material cost.

At times, process simplification can result in the development of an innovative technology that is superior to existing processes. It could be a better batch process or an improved continuous process. Such developments are the most rewarding benefit of the exercise. Each of the above reduces product cost and improves profitability.

Cost reductions are due to improved conversion of raw materials. They are also due to lower conversion cost of a finished product. Therefore, we have to make a continuous effort in each of these areas. A process of continuous improvement allows one to determine, design, and implement the best process for the product.

Cost reductions can also be due to the use of cheaper raw materials. Replacing an existing raw material with a lower-cost raw material has to be part of improving the continuous process effort. Many of the processes, especially active pharmaceutical ingredients, use multiple solvents in a process. Reduction to a single solvent has multiple benefits. Cycle-time improvement improves profitability in two ways: It lowers the conversion cost, and it adds production capacity at minimal or no investment.

If we are able to improve the conversion yield for any product, we are implementing an environmentally sustainable process. Sustainability means "meeting the needs of the present without compromising the ability of future generations to meet their own needs."[1] This is a good definition but I like to modify it to say "exceeding the needs of the present and giving the future generations ability to exceed their own needs." We have to strive to meet a higher goal if we are to be able to meet the expectations of future generations.

In the early seventies when I was at the Illinois EPA working on hydrocarbon emission standards, I recall that during the public hearings the standards were considered to be tough and were challenged by the industry. As the standards became law, industry benefited not only from the savings of raw materials but also from lower conversion costs, since the new regulations led to improvements of the existing processes and the development of innovative processes.

During my tenure at the Illinois EPA I had denied an operating permit to a chemical company because its operation exceeded emission standards of a potential carcinogen. The permit was granted after a reasonable control plan was submitted. About 16 years later I met one of the executives from the same company. He was thankful for our actions, as the company realized a return on investment in a much shorter time than had been anticipated. It had also allowed the company to implement the improvements throughout the operation.

Thirty years later the hydrocarbon emission standards of the early seventies are considered too lax. They were a start. Industry and government collectively have improved the environment. We still have significant opportunities.

Since we are going through a global energy crisis, it would be prudent to figure out how to improve process efficiencies including internal combustion engines so that we can all benefit for the long term. We have to challenge the status quo and strive for better methods. We should not be thinking that it can't be done but looking at what and how it can be done. The current global slump is the best time to rationalize and improve process efficiencies. The goal has to be "innovate, innovate, and innovate."

My purpose in this book is to share my experiences with readers and suggest how we can apply and use our educational and industrial knowledge to simplify and improve process development and manufacturing operations. Readers should not feel that everything has to be done just as I did it but should use this book as a guide for your applications and needs. I have used a few "basic rules of thumb" in my career. This list can be augmented and/or modified to include readers' own experiences. Some of my rules are worth reviewing:

1. There are no failures. Every experiment is a learning experience. These experiences add to our knowledge base and allow us to do a better job.

2. Every dollar we spend has to be earned. My basic rule has been that, under the standard business models, if we are able to increase revenue by $100, we should be able to spend after tax $10. If revenue increase is not the goal but we need to invest to improve the process, then we should save $2 for every $1 to be spent. I have used this benchmark. It may be considered a challenge, but it forces one to be innovative and creative in process selection. I consider this rule as my "breakeven rule." It can be modified to suit individual business needs.

3. We should never hesitate to look outside our business comfort zone for simpler ideas, and we should cross-fertilize. Industry A might be doing the same thing that you are doing but have figured out how to do it better.

4. We should keep in mind that we are dealing with chemicals that many times are alien to our body and our environment. Anything that is alien to our body and environment can be detrimental, as we do not exactly know how it will interact. Thus, it is important that we respect our body and sustain our environment. We will leave a legacy of our deeds.

5. Patents are excellent tools that show what is possible for processes and chemistries. They provide a wealth of information especially for the chemistries and processes that have been and are being invented. Many outline how a process is being executed in the laboratory. They also suggest how the process could be commercialized. Deciphering information can be a challenge but is worth the effort. We need to capitalize on these opportunities.

6. In chemical processes, mass and heat balance are true reflections of the thinking and vision of the developer and implementer or commercializer. They are great educational tools. An actual mass balance reflects the status of the current operation and is a starting place for improvement opportunities.

7. We must document everything we do. It is hard to do, as we are in a hurry to move on to tackle other challenges. Saying "No job is complete without the paperwork" is very apropos.

The above rules are applicable to any manufacturing industry. They are especially applicable to industries that use chemicals. This would include situations where chemicals are reacted to produce a new chemical entity and/or blended for an application to facilitate our lifestyle. In process simplification and operational problem solving, developing a checklist[2] might be helpful. All of us have our mental checklists, but we do not call them a checklist. An organized list on paper should be created and updated with time. It will not only improve the process, but will also result in a safe process and can be used as a training tool.

We humans are the best innovators. If we can go to the moon and come back, we can do almost anything. We want to make our jobs and lives easier so that we can enjoy them. We like to bask in our laziness. After we have enjoyed a good result, we move on to a new challenge.

In order to avoid any stalemate that develops in a project, I have always used a simple methodology of dividing the project at hand into the smallest pieces/steps. A review of all the process steps, allows me to improve each. Let us take an illustration.

Let us assume that a process has five steps: A, B, C, D, and E. We need to review each step individually and collectively. They do not have to be reviewed in the processing sequence. Random order can be used for the review. If process step C can be improved before the other steps, we should implement this improvement to gain its benefit. This not only gives us confidence but also wins over our colleagues. They help us more, as we have facilitated their job. Small wins lead to big wins.

Every manufacturing company has business components that relate directly to the produced products. These include research and development and manufacturing. Process development and process engineering are part of R&D. Maintenance is a critical manufacturing function, needed to keep operations humming all the time. Every other function, such as quality control, shipping, inventory management, sales, accounting, and marketing, are complementary functions of the total business process.

My "rules of thumb" can be applicable to different functions of any manufacturing operation. I will focus on process development, process research, and manufacturing functions.

APPLICATION AREAS

My focus is on specialty and fine chemicals, active pharmaceutical ingredients, paint, paper making, electronic and electroplating chemicals, adhesives, dyes, colors and pigments, and food. These products can be classified in the following two general categories:

- Chemicals that are produced through reactions and may be blended to produce a product
- Chemicals that are blended and applied.

Classification of products in these two general categories is an over-simplification. The fundamentals of chemical engineering and chemistry are applicable, thus there are many commonalities. This not only has allowed me to give a clear and clean view to the challenge at hand but also to cross-fertilize technologies and practices. Since there are commonalities of chemistry and application of engineering principles, compliance with different regulations and safety requirements is simplified.

As we graduate from universities and gain experience in an industry, we generally get labeled as expert in the industry of our first employment and are not considered suitable for other industries. I believe cross-employment has higher value as it offers a different perspective. I would like to illustrate this by the following example.

Organic and inorganic chemicals are reacted to produce a chemical. The created chemical does not know where and how it can be applied and used. If our fundamental knowledge of chemistry and unit processes and operations is strong, we should be able to produce a variety of chemicals for different applications, whether they are a reaction product or a blended product. In certain cases specific knowledge might be needed but that should come in the way of experience of the chemist and engineers.

This is illustrated with the following example. Common salt (sodium chloride) is used in food to enhance taste or treat roads during winter, among many other applications. This does not mean that to mine salt we need personnel with training for each application. However, we do need an engineer who can safely mine and an engineer who can process salt for the different applications.

Similarly, an organic chemical could be used as a flavoring, a UV initiator, a sweetener, a herbicide, an active pharmaceutical ingredient, or an additive by reacting with different chemicals. It is the ingenuity of the chemist and chemical engineer to have an optimal process to produce these chemicals. The chemist and/or engineer should not be labeled a specialist in application/product "B or C," but considered as an innovator who can deliver a quality product safely at the lowest cost. If we classify chemists and engineers based on their past experiences, we are depriving ourselves of their cross-creativity and innovation skills. Commonalities and cross-fertilization provide advantages as they reduce learning and process simplification time. They also bring new thinking in the development of products and processes.

We as chemists and/or chemical engineers need to learn and understand the physical properties of the raw materials and intermediates involved in a reactive and/or blending process. We also need to understand their interaction, reaction chemistry, and kinetics. This knowledge of the chemistry, components, and interactions gives us the capability to control and manipulate the processing conditions. We can be creative and imaginative in improving and developing new processes that will have the following characteristics:

- Economical

- Sustainable

- Quality product.

Knowledge and command of the process variables eliminate any process deviations. This knowledge can allow the development and implementation of continuous processes, which we know are economical and better compared to batch processes.

Use of acronyms such as QBD (quality by design), QBA (quality by analysis), DS (design space), CMC (chemistry, manufacturing, and controls), PAT (process analytical technologies), or any other used by various regulatory agencies to encourage companies to improve their manufacturing technologies become redundant as knowledge of the physical properties and reactions becomes the fundamentals of any chemical-engineering curriculum. Use of these acronyms creates confusion due to variable interpretation.

Imagineering, blue-sky dreaming, and ideation for process and technology enhancements are of considerable value. They lead to innovations. We need to capitalize on "out-thinkers."

Process simplification and innovation are always and will be a "selling" challenge at any company. In September 1973 during a job interview I was asked, "Are you a salesman?" My answer was that as a practicing process engineer I have no experience in sales. I was told that all of us are in sales whether we are a chemist, engineer, or manager, as we are always selling our ideas. The value of this advice has been insurmountable. Many times we are given advice by a colleague and/or in a good book but do not fully adopt it. We should.

REFERENCES

1. http://www.epa.gov/Sustainability/basicinfo.htm. Accessed November 8, 2008.
2. Gawande, Atul. *The Checklist Manifesto: How To Get Things Right.* Metropolitan Books, 2001.

Chapter 2

Process Solutions

Solving process problems is an exhilarating experience but equally a humbling one. Many of us might recall the expression "The thrill of victory and agony of defeat." Solving process problems is a similar challenge. Many may not equate these challenges, but to me there has always been a lesson in how overconfidence can teach you something valuable. Solving a process challenge is a gratifying experience. In solving problems there are no failures but rather learning experiences, which makes you better equipped for the next problem.

Processes, especially continuous processes that operate 24/7 can break down at any time. A solution is needed instantaneously so that production does not suffer. "Process on stream time," whether for a batch or a continuous process, has to be maintained at the set standard; otherwise there will be a cost variance.

Early on in my career, it was suggested that I should be so familiar with the equipment and process that I could visualize each valve and turn of the pipe when called at any place and at any time to solve a problem. It seemed like a difficult task but was one of the best pieces of advice I have had in my career, as it helped me in solving many process challenges. It also assisted me in "daydreaming" process simplifications and innovations.

I developed the following guidelines to review and address every opportunity:

- Understand the materials, chemistry/methodology, and mass balance of the process.

- Spend time on the operating floor and talk with the operating staff. They live with the problems created by the chemists and engineers. They will explain their problems not in engineering terms but practical terms. Understanding their problems makes it easier to solve yours. Many a time there are non-engineering solutions. Common sense works. Explain to the technical and operating personnel the "why, how, and what" of the proposed solution. Incorporate their observations and comments in every change. They are the best support and help in every solution.

- Cross-fertilize technologies.
- Challenge the how and why of the current practice. The solutions will become obvious.
- We have to ask ourselves the question, is this the best I can do for the process? What would I incorporate in the process so that it will have the fewest operational problems?
- When coming up with a solution, never compromise on safety and the environment.
- Always be mindful of the cost of the solution. If you do not see the cost justification for the solution, find a different solution or ask for help.

EXAMPLES OF PROBLEM SOLVING

Mass Balance

Mass balance is an important tool for any manufacturing operation. Its value is significant, as it can be and should be used for monitoring any process on a daily, monthly, and yearly basis. I found it extremely useful, as it allowed me to take preventive and corrective measures if the mass balance deviated from the standard. Mass balance also forms the basis for the factory cost of the product.

The value of mass balance was impressed on me early on when I was earning my undergraduate degree in India. The chemical-engineering curriculum requires practical training at manufacturing plants after the third (junior) year and the final (senior) year before graduation. Each student had to spend four weeks at the allocated sites.

I was at a fertilizer plant and the plant manager asked me to complete the mass balance of a key distillation column. The report was due at the end of my training. It seemed like an easy task until I had to do it. I understood the theory behind the distillation column, but the process operators taught me the practical nuances of various unit processes and unit operations. This showed me for the rest of my life to learn from the operating personnel, as they will teach you the practical, easy, and difficult parts of the process. If one is able to simplify the difficult part of the process, you earn your stripes and you have a friend.

During the four weeks the plant had to change the distillation column packing. I had read about the packed distillation columns but had never seen one from inside. I was asked to enter the column with the full safety paraphernalia. It was a challenge. I saw the guts of a unit operation that normally one does not get to see. Getting your hands dirty helps, as you can easily associate with the plant personnel who have to live with the engineer's good or bad designs.

At the end of my fourth year I was at a plant producing DDT (dichloro-diphenyl-trichloroethane), an insecticide used against malaria. My task during the training was to write the process description, chemistry including chemical structures for each of the raw materials, byproducts, and the product. I also had to prepare a mass balance. It was a challenge but well worth the exercise, as I learned the value of mass balance. For the first time I understood the value of yield and effect of the operating conditions on the process. It taught me how the chemical reactions produce many molecules on a commercial scale, how unit processes differ, and how chemicals can be manipulated to produce a new chemical entity.

This plant also taught me the value and respect of conservation, environmentalism, and human health and safety. The plant was clean but, like any chemical plant, it had its unpleasant sides. On the plant site you could smell the chemicals and feel the DDT dust in the air. None of the obvious places (e.g., dryers) had major leaks, but they were there. One could taste DDT in the water in the cafeteria. It suggested that somehow the process water was getting in the plant water supply, but I was helpless. I also did not understand the ill effects of chemicals. However, they left an indelible impression on me to protect the environment and its impact on human life.

Mass balance includes understanding the materials that are used and produced in the process. This facilitates their manipulation to develop and improve processes.

Monitoring Plant Operations

During my association with the manufacturing plant, we did mass balance for each product every 24 hours. The information included storage tank inventory, in-process material, and changes in the finished-goods inventory for the product. This was a ritual and could be considered overkill, but we lived by these rules. It gave a very good picture of the last 24 hours of the process yield and accounted for all the downtime and any other issues. We tracked our downtime, and every preventive and corrective measure was understood and planned.

Mass balance allowed us to compare and track 24-hour yield against the standard yield. Yields lower than standard were investigated and corrected. This instilled the process of "continuous improvement" in us. It also allowed us to take steps to continuously improve processes to achieve higher yields and improve "on-stream time." In my efforts to solve and/or improve any process challenge, I ask for a mass balance of the process. It is a quick study and reduces the time to address product and process challenges.

Filter Feed Pump

Early in my career in one of my assignments for a process expansion, we had decided on a strategy of stepwise expansion as the process operated 24/7, thereby minimizing downtime. The process had seven unit processes. Based on the mass balance and equipment design criterion, we decided to expand the filtration step first.

A mass balance for the expanded capacity was prepared. Analysis suggested that we would need a higher flow rate at the filter feed pump. I designed the pump, and during a downtime the new pump was installed. The filter was an inline belt filter that used a vacuum to dewater the slurry. As soon as the plant came back on line, the slurry being fed to the filter, instead of having a granular feel, looked like a fine milk shake. Slurry would run down the filter and nothing would filter through. Every effort, including throttling the pump feed valve and lowering the motor rpm, was tried in order to simulate the lower-capacity pump, but nothing worked. More than 24 hours had passed since the installation of the new pump. We checked the design calculations and flow rates and even talked with the pump vendor. The plant could not produce a single pound of the product. Ultimately we decided to reinstall the old pump, and the plant came back on line.

An analysis of the slurry indicated that that the pump impeller also acted as a homogenizer. The slurry became so fine that it would plug the Büchner funnel filter in the laboratory and nothing would filter through. Since we were discharging at atmospheric pressure, ultimately we realized that there was no need to change the pump. We quadrupled the plant capacity but never had to change the pump.

This incident of pump failure always stays in my mind and suggests that a complete problem analysis is necessary. It was a learning experience.

Ester Manufacturing Process

In my job I had moved from process and technology development to manufacturing operations. There I had to live with my process developments, designs, and commercializations and their flaws, whatever they might be. Moving from R&D to manufacturing is a culture change and is worth the challenge. It teaches us how to think and solve problems on the spot. If you get an opportunity like I had to cross-train, you should avail yourself of the opportunity, as it is an enlightening experience.

We produced an ester and used sodium bicarbonate, a mild alkali, as a catalyst for the process. The ester was used to produce a food-grade product, an herbicide intermediate, and a food flavor. Since carbon dioxide was liberated as a reaction byproduct, addition of the solid raw material even with a defoamer was a challenge. Increased amount of defoamer added to the product impurity profile, which interfered with the flavor specifications. We produced about six to eight batches per 24 hours using two existing reactors. We had been using the technology outlined in textbooks. We had tested an alkali in solution instead of sodium bicarbonate in the laboratory. It did give us good yield, but we never commercialized it.

Demand for one of the downstream products of this ester suddenly increased. We needed to double the plant capacity. We did not have the time for a plant expansion. We had to improvise using the existing plant.

We installed a metal deflector at the bottom of the solid feed chute. This minimized wet carbon dioxide to moisten the incoming solids and plug the feed chute. It allowed us to feed the solid as fast as we could feed. Gas evolution did not interfere with the feed, as the metal deflector minimized gas to the dust-collection bag house-mounted at the discharge of the solid feed screw conveyor. Reaction was exothermic. We raised the reactor temperature and controlled it such that it would evaporate some of the alcohol but condense in the reactor top dish and the condenser. Condensing alcohol acted as a scrubber and washed any solids trapped in the gas back to the reactor. This prevented any plugging manifested by lack of pressure buildup in the reactors. We also washed the condenser with the alcohol used in the next batch to clean the condenser tubes, making sure they were clean. We were able to more than double the process capacity and produce 14 to 16 batches per 24 hours to meet customers' needs (see Figure 2.1).

Our ability to increase production using a batch process gave us a number of other ideas, and eventually we converted our batch process to a continuous process. A recent U.S. Patent 7,368,592 (process for the preparation of alkyl N-alkylanthranilate) discusses some of the ideas that we were practicing in the early seventies. Based on my experience the chemistry of this patent can be further simplified, and a yield higher than 80 to 85 percent can be achieved.

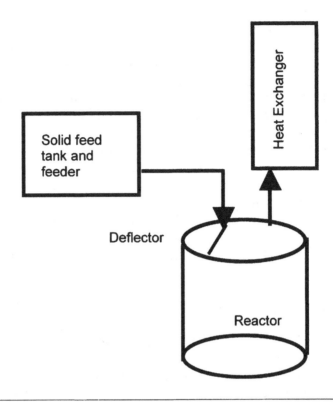

Figure 2.1   Use of metal deflector plate to increase solid feed rate.

Simplification of a Product Solution Process

We produced a product in two forms, a 100 percent solid and a 50 percent version. The product was distilled and the liquid granulated to a solid for sale. To produce the 50 percent solution, the solids were drummed on the first floor and then taken to the third floor and manually dumped in a 4000-gallon tank. Hauling 200 to 220 fiber drums weighing 100 pounds each was an arduous and labor-intensive task. In this process over time we would eventually replace all of the fiber drums. The APHA (American Public Health Association) color of the product was on the high end of the spectrum. Even though the product performance was not affected by the color, high color was perceived to be of lower quality. Due to perceived low quality of the liquid solution, competitors were anxious to enter this lucrative business.

Reviewing the process, we decided to install a three-way valve to divert the distilled liquid to the solution tank. Since the tank was on the load cells, we could control the liquid feed to the dissolution tank. In addition, we were not exposing the distilled liquid to air (as is done in the granulation process); we had a tenfold drop (lack of air oxidation) in the APHA solution. This modification not only improved product color but also reduced our cost and allowed us to increase our total process capacity. We kept competitors at bay. We could simultaneously produce solution and granules if the product demand warranted.

Process "On Stream Time"

In manufacturing it is important to have a high "on stream time (OST)." In plant or process operations the actual time that a unit is operating and producing product is important. OST is a relationship between total available time for manufacturing and the time the product is actually being produced. Table 2.1 gives the generally accepted definition.

For each of our 24/7 operations, the target was to have better than 85 percent OST. This means that we allowed 1260 hours per year for maintenance, repairs, modifications, and cleaning. If we used more than 1260 hours for these chores, our OST would be lower than 85 percent. This might look like a significant number of hours, but in the manufacturing environment it passes very quickly and one has to be very diligent about downtime.

One of our processes, due to its inherent nature of the raw materials and process intermediates, would coat the piping of a continuous process as the time progressed. The pipe diameter would reduce from two inches to a lower size, and eventually flow could stop. Traditionally a planned cleaning was done after the production rate dropped by a certain percentage. Cleaning could take longer than the planned or expected time. This would result in our monthly OST dropping well below 85 percent. The average production rate would drop to about 2500 to 2800 pounds per hour, below our design capacity of 3000 pounds per hour.

Our goal was to improve the OST and production rate. I met with the operating personnel to discuss every option to improve the operation. Cleaning was our largest downtime contributor. We decide that instead of cleaning the process piping as had been done in the past, we would clean every 48 to 60 hours.

Table 2.1 On-stream time definition.

Available time/year, hrs.	85% OST
$350 \times 24 = 8400$ hours	$= 8400 \times 0.85 = 7140$ hours

Our idea was to minimize the buildup by giving the equipment a quick clean. We documented the needed cleaning time and the subsequent operating rates. After a few weeks we reviewed our operating rates and OST data. We were delighted to see that our OST jumped to more than 85 percent, and the operating rate increased to about 3200 pounds per hour, better than the designed rate.

This improvement allowed us to produce above our budgeted rates. It was also very timely, as the demand for a downstream product increased, consuming every pound we could produce. We achieved our objective of meeting customer needs without any investment and also increased our profitability. This gave us the flexibility to shift our production from 24/7 to 24/5 if it was necessary. Since we were overabsorbed on our variances, we did not lay off anyone and used the spare personnel for other needed tasks. It was a morale booster for the site.

The concept of earlier than planned cleaning can be applied to any process where piping and/or equipment cleaning is necessary. Membrane cleaning, such as in biopharmaceuticals, ultra-filtration, and the like, are good examples.

Inventory Management and Process Repeatability

Since I had moved from process development and R&D to manufacturing and I did not know any better, I had no preconceived notion of norms for the inventory. My goal was to smooth out the monthly production so that we could minimize the end-of-the-month sales rush.

I also had the luxury of managing the raw material, in-process, and finished goods inventory to as low a level as I chose so long as I could meet my production rates and the sales demand. I considered inventories to be an "operational evil" of any process. In order to minimize this evil, my goal was to minimize the inventories.

To achieve this goal, I made every effort to understand our processes, their pitfalls, and their break points. This exercise was helpful, as we could repeat our operational mistakes and correct them. During my operation days (1979–1984) our goals were to operate with close to minimum inventory and produce quality the first time (six sigma in today's terms).

Since we had excellent control of our processes and "on-stream time"—i.e., we could produce and supply customers as they placed orders—our orders leveled, and this gave us an opportunity to minimize the inventories of raw materials, in-process materials, and finished goods. With the help of our purchasing and other support staff we operated by maintaining inventories in the range of one to two days for any of the raw materials, one day for in-process materials, and two days of regularly scheduled finished products. We were able to achieve these goals. The cash flow of the profit centers was excellent.

Sampling and Size

In every manufacturing operation, it is necessary to test the intermediate samples at prescribed points and also check the final product sample. Sample size for any analysis used to be a quart (liter) sample. Most of the time the samples were disgarded after testing and would end up in waste. In those days environmental laws were very lax compared to today's standards. In our operation, change from a liter sample to a 250 ml sample was met with significant internal resistance. It was like going upwind. To get around this dilemma, the plant decided to have sample accumulation drums. This encouraged recycling and waste minimization. Eventually everyone understood the value of small samples, and the plant settled on 250 ml or less.

The concept of minimum size samples has stuck with me. It is not only an effort to minimize waste but also to improve yield. Leftover samples should be recycled if possible, and/or sample size should be minimized. Samples have significant value when the products are expensive. Reduction of the sample size is an effort to green the process. In order to have a representative sample, a careful review of the sample point, method, and size is necessary. Flush valves are available, and they should be used wherever possible. They are worth the investment.

Recently a process audit at a client business indicated that four one-quart samples were being taken during the batch. The samples after testing were being discarded as waste. After discussion with the laboratory personnel of the contract manufacturing organization (CMO) it was indicated that they did not need a quart sample per test. They were just following the established procedures. It turned out that 100 ml per sample was sufficient. The reduction of sample size and recycling saved them in excess of $600,000 per year.

Sample taking is a simple exercise but a critical place where yield loss can occur with a small batch and an expensive product. Many times this is overlooked. Better sampling valves have been developed and are commercially available. They should be considered.

Antioxidant Production

An antioxidant was produced at a plant by reacting a substituted phenol with formaldehyde in presence of a catalyst. The process as defined had a simple chemistry and a reasonable stoichiometry, suggesting little more than two moles of formaldehyde should be used for the process.

However, the process produced quality product randomly—i.e., we could never predict which batch would meet specifications. The batch charges to the reactor were weighed and meters were calibrated to decipher the cause of our inability to produce quality product. However, the problem persisted. If the product was not going to meet specifications, it would only show up at the filtration step, as the centrifuging would be a problem. With the provided stoichiometry it would take about five 24-hour days to produce two to three batches per week.

Since the plant was purchased from another company, there was not much information available about the chemistry and the process. Our only choice was to reinvent the chemistry in the lab and compare it to the actual chemistry being practiced in the plant. Based on mass balance, about 20 percent excess formaldehyde was being used per mole of the key raw material. Since the theoretical chemistry asked for two moles of formaldehyde, experiments using 5 to 8 percent excess formaldehyde were tested in the lab. The laboratory batch met specifications. The process in the laboratory was very easy and clean. We reconfirmed the batch in the lab with excellent results.

Since the customer orders needed to be filled and we had enough confidence in two to three lab experiments, the process was commercialized. Formaldehyde needed for the new process was less than the handed-down recipe. The process yield was about 90 percent, and the product quality was higher than the established specifications. The plant was able to produce five batches per week compared to the earlier two to three batches per week. Analysis of the product produced using the new stoichiometry suggested that the excess formaldehyde produced by-products that caused filtration problems and also resulted in off-spec product.

Mass balance and attention to stoichiometry not only solved the problem but also allowed us to produce more product in less time, adding to the profitability of the plant. As I have indicated earlier, monitoring mass balance keeps track of the process yield. In addition, it monitors the pulse of the process. Operational problems will show up and should be resolved immediately.

The acquired plant produced another antioxidant, which was one of the major products for the site. However, we were losing more money than can be imagined. Our task was to fix the technology and the operation so that we could return to profitability.

Our first task was to understand the chemistry and the process and create a mass balance. Creating a mass balance was a challenge, as the necessary systems and support were limited at this acquired facility. This was complicated by the fact that some of the residues were being sent out for recovery and disposal. Mass balance should normally close if one has control of the process and an understanding of every process stream. Closing the mass balance is accounting for the process and establishing an actual yield. With limited resources, we developed a quick analysis for each process stream. A large gap existed between the theoretical and the actual yield.

To account for the gap, information (capacity, configuration, and calibration) about every tank was compiled and chemical analysis of every major stream completed. Under the established practices, we recycled process streams to recover raw material. The distillation bottoms were being sent for disposal. An analysis of the material being sent for disposal indicated that the waste stream was 90 percent product. Since it was pitch black, not much attention was being paid to it. The company was paying money to get rid of it. Disposal was immediately stopped and measures put in place to recover the product in-house.

Standards for composition of the waste stream were established. Every step of the process had to be reinvented. A process operating and technical manual was prepared. This was followed for everyday operation. The profit center turned profitable. Again, my emphasis is on monitoring the mass balance, knowing the materials, and understanding the chemistry and how it is being executed. It will point to operation problems and assist in solving them.

Novolac Resin

The plant produced many novolac resins. One resin had the largest volume. The reaction was simple condensation chemistry, taking the reaction mass to a certain ring and ball softening point. To improve profitability, the plant needed to reduce the cost of every product. Since this product was one of the largest in volume and simplest to produce, we decided to review its process.

Based on the discussion with the resin technologists, I concluded that the plant had scaled up the laboratory synthesis and comfortably produced the resin as developed. This was also the industry tradition for such resins and other similar products. Laboratory equipment has limitations. Things that can be accomplished in a commercial plant cannot be tried and/or replicated in the lab.

The plant took one eight-hour shift to complete a batch and the same time to flake it. Moving from a batch to a continuous process or significantly reducing the cycle time was the only hope to reduce costs. Reaction process and flaking operation were reviewed separately. Most of the time such segmentation of a process is very good, as it facilitates problem solving and simplification. Through some quick tests and heat-transfer calculations we determined that the flaker had the capacity to take the additional load. If we needed additional cooling, it could be increased with a chilled-water system.

Experts warned us about using a continuous process, as a runaway reaction could jell the reactor and be unsafe. Since we had experienced few jelled reactors, the idea of continuous process was temporarily shelved. It was decided to take a stepwise approach to reduce the current batch cycle time. Every hour reduction in the batch cycle time reduced our cost and improved our throughput.

We divided the batch into its components:

- Charging the reactor
- Reaction
- Dehydration
- Emptying the reactor

This breakdown was used to determine how we could reduce time for each individual step. The first and last steps were pump-related and the easiest to improve. Pumps had capacity, and we were able to reduce the cycle time to less half the time taken by these two steps. Having accomplished pumping time reduction, the next step was to determine how we could reduce the reaction time.

In this particular process, it was obvious that the reaction was very fast. The longest process step was the removal of the water, a reaction byproduct, until we reached a designated set point. Our chemists were very familiar with the steps involved. We used equipment capabilities and heat transfer to improve the reaction rate and reduce the processing time.

Overall, we were able to reduce the cycle time from eight hours per batch to about 3 to 3.5 hours per batch. We comfortably doubled our production capacity without spending any money. We used to operate 24/7. By doubling the capacity, we could out-produce in 24/5 and supply the product to the customers. The resin production turned from a loss leader to a winner.

Based on our experience, it became obvious to us that there is a definite possibility to convert the batch process for this novolac resin to a continuous process. On a case-by-case basis the thinking can be extended to other novolac and alkyd resins. Careful design and processing controls can deliver much superior results than the traditional processes.

Alkyd Resins

Later on I had an opportunity to review various alkyd resin processes. Based on my review of the alkyd resin processing methods and their similarities to novolac resin chemistries, there are definite opportunities to shorten the batch-cycle time. Some processes can be converted from a batch process to a continuous process. I am sure readers are familiar with the benefits of continuous processes. It has to be a case-by-case decision. Traditional thinking has been that the alkyd resins cannot be continuously manufactured. My belief is that such thinking persists because no one has thought about it and explored it.

Paint Making

In water- or solvent-based paint manufacturing chemicals are dispersed in formulation solvents using traditional high-speed dispersers. A certain particle size is achieved, and appropriate additives are added to produce a paint. Colors are added to shade the paint if necessary. Traditionally, high-speed dispersers are the workhorse of the dispersion process. Not a lot of effort has been made to explore alternate dispersion methods on a commercial scale.

At a client company a product was reduced to a certain size using a high-speed disperser (HSD), and further size reduction was needed. The client used a ball/sand mill. These mills have a certain function and they fulfill it. Due to the mill volume, a large amount of solvent is needed for cleaning, and it may or may not be reused for the manufacture of the coating. If the solvent is recovered and recycled, it generates sludge that has to be disposed of. In addition, these mills take a long time to produce the desired product. In today's environmentally conscious world it is helpful if higher efficiency and lower polluting technologies can be used.

The client needed a technology that is commercially produced, economical, and less polluting and performs equally to or better than the ball/sand mill. I recommended commercially available high-shear rotor/stator technology, which I had used. This technology was tested on the client's product. It was successful. The gains from the technology have been significant. The client can now finish the batch in less than two hours, including cleaning, compared to the six hours needed to clean the ball/mill. Since the cavity volume of the rotor/stator is much smaller than the mill, a significantly lower volume of solvent is needed. As the cleaning-solvent volume is low, it becomes part of the batch, thereby eliminating solvent recovery. Since the residence time is low, less heat is generated, and solvent loss due to evaporation is minimized. Compared to high-speed dispersers, especially for solvent-based products, this technology is an "environmentally sustainable" technology.

The client has successfully more than doubled capacity with less than 25 percent of the investment cost of a comparable ball/sand mill. Since the rotor/stator can act as a pump, it can recirculate the material along with reducing the size. When used along with the regular high-speed disperser, it can further reduce the batch cycle time.

Rotor/stator machines can be portable to serve multiple tanks. This can reduce the total investment. These machines are versatile and should be evaluated along with conventional technologies for two important reasons: cost due to reduced time to produce a batch and green process due to minimal solvent use, potentially eliminating waste.

This is not to say that the ball/mill or other similar equipment should not be used for its intended purpose but that other technologies, used to produce similar products, should be evaluated and considered. Cross-fertilization of technologies can be economical and beneficial.

Use of these high-shear rotor/stator machines in the continuous manufacture of paint will be discussed in another chapter. Use of rotor/stator technology is similar to use of microreactors in a chemical reaction. They are versatile and require less investment but their return on investment must be carefully evaluated.

Continuous-Process Latex Polymer

Latex polymers are conventionally produced using a batch process. My company had developed a proprietary latex polymer for an application. Since the polymer was a proprietary, it was decided to produce it internally. The latex could have been produced using a batch process. It was decided to produce it via a continuous process. We were successfully able to produce the polymer by finessing the sequence of addition with a back-mix reactor. This was achieved by understanding the physical properties of the reactants and the reaction chemistry.

The particle size of the latex was in a narrow range and better than a comparable batch process. The monomer in the finished polymer was below the desired level, thus we did not have to strip it. If the polymer were to be produced using a batch process, the unreacted monomer would have to be stripped. In addition, the product from the batch process had a wide particle-size distribution. Some of the rationale and the development method for a continuous latex process are discussed in another chapter.

Waste Paint

Paint companies, in addition to inventories of off-spec paint, accumulate significant quantities of waste paint. Waste paint is an accumulation of customer returns, poor color matching, outdated paint, and damage at the warehouse. Appropriate disposition of this paint has been and continues to be a challenge. However, beyond collection challenges, it requires significant imagination to determine where and how it can be used.

Two solutions have been prevalent. One is to blend the same solvent-type paint. The blended paints are a light shade of gray or green. The other solution is to solidify the paint; it becomes part of the household garbage ending up at a landfill. Attempts have been made to use these two methods, but still the waste-paint challenge exists as either method requires effort.

There are two other alternatives for the waste paint, especially the water-based latex. It can be used as a raw material for new batches of paint. This is feasible but requires constant attention, as the raw materials (waste paint) can vary from day to day. Since this effort is a constant challenge compared to use of virgin raw materials, interest wanes and the process is stopped. We developed a methodology that was commercialized, but it is a continuous challenge and is not for the faint-hearted. It is easier to justify why it cannot be done rather than how it can be a success. In addition, environmental pressures and housekeeping can be a roadblock to its success. Kelley-Moore, a West Coast company, has commercialized 16 colors of paint that contain a minimum of 50 percent post-consumer latex-based paint. Metro Paint in Oregon also produces recycled paints. These products illustrate that paint recycling is feasible. Companies have to select the route that is workable for their operations.

The other alternative is to convert the waste (recycled) paint into a raw material that can be used to produce other products. This avenue requires significant creativity. In order to develop these uses, the paint company has to become a waste-paint processor, converting the material to saleable raw materials. However, the current EPA regulations do not promote such initiatives, since the paint company could be classified as a waste processor and would have to contend with a different set of rules.

Avenues such as using the waste paint as an additive in papermaking, wallboard binder, solid powder (produced by spray drying), and/or concrete do exist. These avenues have an economic advantage, as waste/recycled paint has significantly lower monetary value than fresh raw materials. This is a value proposition, as it can lead to higher profits. This effort requires creating a new business within a company. I have explored some of these avenues.

Developing these applications requires an understanding of the industries where the raw materials are used. Understanding raw materials and their functionality is necessary to explore nontraditional avenues. It also requires open-minded and creative formulators who have the funding to explore these avenues. Use of paint as an alternate raw material for paper will be discussed in another chapter.

At times paint makers have found it is economical to ship the paint to third-world countries for sale. With rising fuel costs shipping to the third-world countries is becoming an expensive option.

Wastewater Treatment

A client was managing a wastewater treatment facility for an aluminum factory. It had to treat a minimum volume per day so that the plant could keep operating. However, there was significant fluctuation in the volume of water being treated every day. At times the aluminum production had to be curtailed until the wastewater treatment was able to catch up. This was not economical for the aluminum producer.

Before corrective actions could be implemented, a review of the mass balance and procedures being used at the treatment facility was necessary. There were set procedures that needed to be followed to sustain constant operation. However, all shifts modified treatment procedures to ensure high volume for their shift. Constant modifications without any practical basis led to considerable deviation of the effluent treatment. It also resulted in upsetting the system equilibrium and in significant variation above the set effluent limits. In order to stay below the EPA effluent limits, production would often have to be curtailed. This was not acceptable to the factory.

An operating procedure as outlined in Chapter 8 was prepared and implemented. To ensure that there was no deviation from the set procedures necessary, metering pumps that pumped the treatment chemicals at the prescribed rate were installed. Within days of the implementation of the procedures, the facility was able to increase its production by more than 15 percent, as the wastewater treatment facility could handle the load and had extra capacity.

Adherence to set procedures and methods that are based on process knowledge and mass balance are important for any operation to work smoothly. Unauthorized deviation from set procedures can upset the process and have undesirable results.

Electroplating Chemicals

Many electroplating chemicals are a blend of chemicals. Some of the chemicals are manufactured by chemical reaction to produce a complex that becomes the protective coating. As I have indicated, patents are a magnificent knowledge base and are worth harnessing. They can be used to jumpstart a product-development process and reduce time to market. It can lead to the development of a better product.

For the development of an electroplating chemical, we had a limited time. Using the strategy outlined above, we were able to develop a commercially successful product meeting the upcoming regulatory requirements within weeks rather than months.

In another instance, a relatively simple reaction caused a violent exotherm. This was safely handled by reverse addition of the chemicals. In addition to reverse addition, we were able to use higher specific heat along with controlled feed rate to absorb the heat of reaction and reduce the temperature rise. In this instance, we manipulated the physical properties of the chemicals to control the reaction process. Chemistry was not changed but rather the order of addition, allowing a safe process. This change also eliminated an expensive investment and reduced pollutant emissions.

Understanding and managing physical properties of various chemicals can simplify many reactions. This is further discussed in Chapter 7.

Specialty Surfactant Manufacture

A specialty chemical company had simple but exotic chemistry for multiple products using similar molecules. Products were commercialized. However, the batch-cycle times were extraordinarily long. Some of the processing steps were scale-ups of the laboratory practices. The products had excellent promise to replace environmentally unsafe products on the market. It was necessary to institute a competitive process.

Based on a review of the chemistry, raw materials, and the lab process, it became obvious that the chemistries and the manufacturing steps needed to be executed in different ways. The client had an existing pilot plant that was not used for specialty chemicals. Processes had to be accommodated in the existing equipment with minimal investment. Batch-cycle time had to be reduced from the initial trials so that major investment in a new plant could be postponed until the market developed.

A careful review of the processes suggested that the existing equipment could be used, but the chemistry would have to be executed differently. Minor equipment investment was needed. The order of addition of the reactants was changed. The reaction temperature was raised to improve the kinetics. By improving the reaction kinetics and using physical properties that included densities, heat of reaction, viscosities, specific heat, and azeotrope characteristics of various chemicals to our advantage, we were able to simplify the process. We were able to postpone investment in a brand-new plant for the foreseeable future.

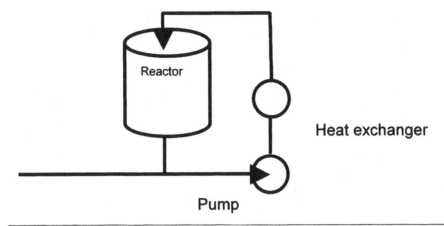

Figure 2.2 Use of external heat exchanger.

In one particular case the reaction with a certain chemical was highly exothermic so that the localized heat would impart color to the product. In order to prevent the color formation, the method of reactant addition was modified. Figure 2.2 shows a simplified version of how the reaction exotherm was controlled. Reactant was added in the pump cavity, allowing vigorous mixing and controlling the exotherm using an external heat exchanger. This method improved the reaction time.

Process and equipment for the products are configured such that the client is able to produce electronic-grade materials using equipment that was never designed for such high purity. It is possible due to simplification of the process.

A microreactor (nano version) of a plate and frame heat exchanger could also be used. However, based on economics the suggested solution was cheaper and could be installed in a much shorter time. We should always be creative and imaginative.

Decanter

If two immiscible solvents are used in a reaction, phase separation is a natural part of the process. Azeotropic distillation, in addition to decanters or centrifuges, is a common method. Phase separation with a decanter does not use energy. Azeotropes, besides using energy, still require phase separation.

Decanters are the least understood of the types of equipment used to separate two liquids of different densities. Most of the time chemists and engineers think of them as a separatory funnel. I have discussed their design details in another chapter. They are the simplest type of equipment, exploiting the density of the involved chemicals/solutions.

I can go on describing many cases in which we made simplifications and improved the phase separation. I have given an example of one case. In one process design a significantly large engineering consulting company had designed a 100-gallon phase separator with all sorts of instrumentation to control flow rates and phases. The uninstalled cost of the phase separator was close to $100,000. The smallest decanter, due to all the ports that would have been needed, was about 30 gallons and cost more than a reactor of similar size.

I reviewed the physical properties of the solvents used, and, based on their density measurements, the phases separated instantaneously when the agitation stopped. A five-gallon glass decanter based on hydrostatic pressure balance that cost less than $1000 was installed. It improved the processing time multifold. It is totally maintenance-free equipment.

If for some reason the phase separation fails, it is an indication that the process chemistry was not correct and that there are process problems. Imagination, simplicity, and measured exploitation of the physical properties of the chemicals are the key to process simplification.

Active Pharmaceutical Ingredients

We have to recognize that the small-molecule active pharmaceutical ingredients (API) are specialty/fine chemicals that have disease-curing value. These API are formulated with excipients to facilitate dispensing. Known unit processes and unit operations that are taught in the chemical-engineering curriculum are applied to manufacture and formulate API. There are many opportunities to improve their processing, quality, and costs and thus profitability.

It would be beneficial to review the reasons for the current state of pharmaceutical development and manufacturing processes. Regulations were created to ensure that consistent-quality drugs are delivered. They also need to be safe for human and animal consumption.

Since safety and product repeatability are the priorities set by various regulatory agencies, innovation in process development and manufacturing technologies, necessary in any business process, is lacking in pharmaceuticals. This is due to the expense involved in re-approval of the products. Limited patent life and speed to market also prevent manufacturing and/or business innovation if a molecule demonstrates disease-curing value and efficacy in clinical trials. Other innovation deterrents are patent monopoly, as no one else can produce the same drug during the life of the patent, thus the pharmaceutical company can achieve its profit margin. The premise here is that there is no need to invest in a product that delivers your financial objectives and will not be produced company after the patents expire.

Generic companies in recent years have increasingly challenged patents, but usually these lawsuits result in a win-win for the brand and the generic company. The consumer does not benefit much, as no one innovates the manufacturing technologies to lower costs. The potential to lower costs does exist, but it is not a priority.

Manufacturing methods and techniques developed in the laboratory are scaled up to commercial scale. If the laboratory process is inefficient, complex, low in conversion yield, and even environmentally "not sustainable," it is still commercialized. Any associated costs are passed on to the consumer. Since we all want to extend our lives, we are willing to pay the demanded price. Drugs are sold at the highest price we are willing to pay.

After the patents related to the drugs expire, selling prices do drop but not precipitously if they are commoditized. This is again due to the highest price we are willing to pay to extend our lives.

Manufacturing process inefficiencies are accepted. They are due to less than optimum use of process equipment and less than optimum conversion yields of raw materials. These in turn are due to a lack of complete understanding of the interaction of chemicals used to produce API and formulate a drug.

Lack of process understanding can result in quality variation in manufacturing for every intermediate. Since the intermediates are checked for quality when the molecule is being developed or is being formulated in the laboratory, these methods are translated and scaled up in commercial operations. Quality analysis of intermediates and assurance that they meet intermediate and final specifications of API and drugs is called quality by analysis (QBA).

Due to drug safety and performance repeatability requirements, regulations do not promote innovation. Regulatory agencies are not responsible for manufacturing innovation. They would like to see process innovation. However, the fear of re-approval and the associated costs are a self-inflicted industry roadblock. It would be worth exploring the state of API manufacture and reviewing the opportunities.

In the production of many APIs, it is customary to isolate and purify the intermediate and the final products. The methodology of producing quality product has met regulatory guidelines. Thus, quality by design (QBD) manufacturing technologies have not been necessary.

Since the laboratory development is a batch process, most of the API and formulated drugs on commercial scale are also produced using a batch process. The combination of QBA methods, batch processing, cleaning the equipment after every step, use of multiple solvents, and other factors results in an inefficient commercial process.

QBA has stifled innovation and improvements in manufacturing technologies for API as well as drug formulation. Due to QBA methodology most of the API are produced by batch process even if a continuous process can be used.

QBA also creates challenges from a business perspective. Since the batch cycle times are extended, total investment in the process equipment, infrastructure, and inventories is at much higher levels compared to a similar specialty chemical plant manufacturing products that do not have a disease-curing value. Synthesis yield of API is generally lower than a specialty chemical. In addition, intermediate isolation further lowers the yields. All these steps extend the batch cycle times. Lower yields increase waste-treatment load and increase energy use. A chemical produced using methods that are used to produce an API will have a much higher cost compared to the same chemical produced by methods used to produce specialty chemicals even if the quality is exactly the same.

QBA, QBD, PAT (process analytical technologies), DS (design space, another definition of process operation parameters), and CMC (chemistry, manufacturing, and controls) are the recent buzzwords in the manufacture of API and drug formulations. It is a general belief that PAT will solve manufacturing problems, streamline the manufacturing process, and produce quality product. Unless the process chemistries and the interaction of materials are completely understood and translated to appropriate processes, unit operations will not produce quality product. Process analytical technologies do not correct the problem. They just indicate that there is a problem.

I have come across situations time after time where the chemistry as practiced can be simplified. No one has challenged it during development or practice. Hierarchy-centric organizations and cultures, in addition to all the above reasons, have stalled process and manufacturing innovations. Whatever method and/or process is developed for an API and its formulation, it is not challenged and/or changed, as it will challenge the intellect of the person (mostly the inventor or the entrepreneur in the developing countries and the API inventor in developed countries) who has developed the process. In addition, there is a fear that change might delay commercialization or that an existing product and/or process might have to be reapproved though the regulatory quagmire.

Since 2005 the global pharmaceutical market has been in an upheaval. Many of the patents of the big-selling drugs from major pharmaceuticals companies will expire by 2012, and their pipeline is sputtering. In addition, generic pharmaceutical companies have been challenging the patents. This, along with the regulatory agencies asking for process and technology improvements has presented an opportunity for every API manufacturer and formulator to improve. With the reality of declining revenues and profits and declining drug discoveries, the brand (ethical) pharmaceutical companies should expend effort to improve API manufacturing and formulation technologies.

Following are a few examples in which understanding the process chemistry and managing it improved process productivity and reduced costs without impacting the quality.

In a recent case for an API, an intermediate was isolated from a solvent, dried, and dissolved in the same solvent for the next process step. The next process step could be done without isolation and without changing the product quality. Forgoing the isolation, drying, and subsequent dissolution steps decreased the batch cycle time by about 20 percent.

In another case a reactant was being used in excess. Besides acting as a reactant, this chemical also acts as a solvent. Due to the kinetics of the process and the excess the process step yield was about 50 to 55 percent due to the formation of byproducts. Excess chemical also reduced productivity of the reaction step. The reason for the low yield is that the reaction product is further reacting with the solvent to produce undesirable products, thus lowering the yield.

An alternate method was developed whereby the addition sequence of chemicals was altered and the reaction product was immediately removed. In this method any excess of the solvent acting as a reactant was not needed. The yield and productivity improved.

In each of the above cases, chemistry was not altered but rather executed differently to take advantage of the physical properties and reaction behavior of the chemicals. The list is long and I can go on with numerous cases, but pharmaceuticals, which are specialty chemicals, need to apply basic fundamentals of chemistry (organic, physical, inorganic) and chemical engineering to simplify their manufacturing methods and move from laboratory technology to commercial technologies that are being practiced in manufacturing similar molecules. These are discussed in other chapters and references 1, 2, 3, 4, 5, 6, 7, 8, 9, 10, 11, 12, 13, 14, 15.

API Intermediates and Pollution

In recent years many of the organic intermediates necessary for the production of many active pharmaceutical ingredients and many APIs are being produced in China and India. This has cost significant business revenue for the companies of Europe and North America that were the producers of these chemicals. The general complaint has been that the companies in the third world do not have to comply with the environmental standards that are being applied in the developed countries. This conjecture might be true, but the companies are complying with the prevailing laws of their countries. The real reason for the production of these intermediates and APIs outside the developed countries is that the processes for the manufacture of these chemicals are inefficient, resulting in poor yields. Poor yields of these APIs demand high investment to comply with the prevailing laws of the developed countries. Since the companies decided against the investment to develop better processes, most of these intermediates and API are produced outside Europe and North America. Opportunities are still available to have safer and more economical processes.

The ecological impact of APIs and their intermediates and drugs is a cause for alarm.[16, 17, 18] Companies might be in compliance with the prevailing environmental laws, but the toxicological impact of many of the intermediates and active ingredients is not known. Recent news reports are not encouraging. Ecotoxicity is a cause of concern. Again, one way to reduce this is to have efficient manufacturing processes.

SOLVING PROCESS PROBLEMS

Following are a few examples of the problems chemical and coating companies have and their solutions.[19] In each of the provided solutions my intent has been to minimize investment and downtime.

Maintaining Pump Prime

PROBLEM

While pumping ammonia to our unit in eastern Indiana, the centrifugal pump loses prime for no apparent reason. It then takes several attempts to get the pump primed again. The problem seems to occur more often in the summer but has occurred in every month of the year. The ammonia is stored in 10,000-gallon insulated saddle tanks with a positive head of at least 10 feet when the tanks are nearly empty. The typical flow rate is between 5 and 10 gpm. The pump has a minimum flow requirement of 5 gpm, so there is a minimum flow line with a control valve to maintain the flow. What could be causing this problem?

Degas the Ammonia

We used to have a very similar problem when we were pumping an aqueous reactant solution that had sulfur dioxide as one of the reactants. The pump would cavitate, and we would lose all of the flow, which would shut down the process. We determined that this was happening because of the dissolved gas in the liquid, thereby lowering its density to , less than the minimum NPSH (net positive suction head). We would shut down the process and let the liquid degas until the density became normal and then restart, which often took several attempts. We were able to solve the problem by maintaining sufficient head and making sure that there was no dissolved gas in the liquid.

Take out Torque

Problem

We are experiencing high torquing problems and an increase in material temperature in our six-foot-diameter ribbon blender. We mix a dry, coarse ingredient with oil and then add a fine powder and mix for about 30 minutes. We feed the coarser material into one end of the blender and add the powder to another area. The oil is sprayed into the mixture using fan sprays. About 20 minutes into the mixing process, the torque increases substantially and the bulk mixture temperature rises to about 90 degrees F. What could be causing the torquing problem? Will changing the mixture reduce the overload?

Slowly add Fine Powder

I believe a simple change may solve the problem of higher torque and heat. Assuming that there is no reaction taking place between the oil and solids, I would spray the oil on the coarse powder and run the blender until the oil thoroughly coats it. Once that is done, I would slowly add the fine powder so that it starts to adhere and coats the oil-coated coarse particles.

Stop Solvent Losses

Problem

We make adhesives by batch-processing solvents and about 1500 pounds of solids in 3000-gallon tanks. Our base solvents are hexane, acetone, and isohexane. The formula calls for a three-hour mix cycle on high speed to dissolve the solids; at this time we lose about 15 percent of the solvent volume. Although the tanks have fixed lids, they are not pressure vessels. Is there a way to reduce the solvent losses by altering our mixing procedure or tank configuration? Is there any other way?

Cut Back the Mixing Time

The description suggests that it takes three hours to dissolve the solids in the solvent mixture, which is part of the problem. Review the dissolution process to see if the time can be reduced, which will result in lower solvent losses.

In addition to reducing the cycle time, it is necessary to keep the batch temperature as low as possible. This can be achieved by installing a heat exchanger that will maintain the batch temperature close to ambient, thereby minimizing solvent loss.

Two types of heat exchangers can be used to control the temperature. Options are an immersion-plate heat exchanger in the tank or a jacket or external-plate heat exchanger attached to the straight side of the reactor. Water can be used as the coolant in both cases.

If the batch tank does not have any baffles, a properly designed immersion-plate heat exchanger will also act as such and improve the dissolution time. If a jacket or external-plate heat exchanger is used and the tank has no baffles, you should consider installing some.

You will have to evaluate the economics of each temperature-control method to select the best option for your plant.

Choose a Pump Wisely

Problem

We need to pump a reactor effluent stream. The stream is slurry, which is highly aerated by an agitator in the reactor. We estimate that 10 to 15 percent of the pump suction volume is gas. The total volume at the pump suction is about 1200 gpm (including gas) at a head of around 70 psi (200 feet). Which kind of pump is best for this service?

Try a Rotary Disk Pump

You should consider using a rotary disk pump; it works. This type of pump uses a series of rotating parallel plates to create a boundary layer and viscous drag force. The plates allow the pump to transfer the liquid as needed.

Know When the End is Near

PROBLEM

At our coatings plants, we produce many types of synthetic resins, and for each type we have several recipes. Is there an easier, more reliable way to monitor the product viscosity and determine the end of the reaction other than the classical manual sampling and laboratory testing?

USE AN INLINE VISCOSITY MEASURING DEVICE

An inline viscosity measurement will eliminate trips to the lab and has three significant benefits:

- Reduction in batch cycle time due to elimination of offline testing time
- Consistent and improved product quality since the end point of each batch will be the same
- Elimination of waste.

There are a few inline viscometers available commercially. Since you want to check the reaction end point, it would be advisable to have a pump-around system for the analyzer, which can be solvent-washed between batches. You will also need to heat-trace and insulate the pump-around piping to prevent any solidification of the resin.

Vanquish Material Variations

PROBLEM

We are experiencing variations in both our raw materials and finished product. We chemically analyze the raw materials using an online vertical analyzer, but the data from the raw materials do not correspond to those for the finished product. We have experienced hangups and segregation in our feed silos, which might be exacerbated when the raw materials are mixed and then transported on a belt conveyor. We suspect the variation in analyzer readings could be due to increased moisture in the raw materials after leaving the feed silos, but we don't know why. Is the difference in readings due to segregation or to moisture?

CHOOSE MONITORING POINT CAREFULLY

I would suggest the company take the moisture reading after the material has been mixed and before it gets to the belt conveyor. The problem may be caused by any of the following or combination of reasons:

- Variations in moisture content of each of the incoming raw materials
- Hygroscopic nature of the raw materials
- Materials of different moisture content hanging up in silos and then mixing.

The mixed raw materials have to meet an acceptable specification. Otherwise there will be a variation in finished goods. Use of a moisture analyzer after mixing raw materials will also allow tightening the virgin raw-material specifications and improving the total process.

Fend off Reboiler Fouling

PROBLEM

We recover ammonia by distilling an aqueous solution that contains less than two percent organics (fatty oils). The column is under pressure, and has a forced-circulation reboiler that uses a heat transfer fluid for heating. The reboiler has been experiencing fouling on the process side from the organics in the feed; it is worst when the system is restarted after maintenance. The fouling rate is not very predictable; sometimes the reboiler runs for a while before the heat transfer drops off. What is the best way to remove the oil from the feed?

REMOVE THE FATTY OIL FIRST

Fatty oils have a tendency to oxidize to fatty acids under high temperature and pressure. It is possible that this is the case. This can be confirmed by checking the pH. If the pH of the liquid in the reboiler is below seven, there is acid in the reboiler. However, due to ammonia, the pH should be alkaline or close to neutral.

If there is acid in the reboiler, it will attack the metal, especially if it is carbon steel. If the oil phase is not removed, its concentration will increase, thereby adding to the possibility of corrosion.

An option is to remove the fatty oils before the aqueous solution is fed to the recovery unit. This could be done via physical means such as a centrifuge if the oil would separate or by chemical treatment (coalescing agents) and then some form of physical separation.

Prevent Heat Exchanger Plugging

PROBLEM

Our solvent recovery system uses a batch evaporator to concentrate the mother liquor from a crystallization process. The evaporator is a falling-film unit with a recirculation pump and a shell-and-tube heat exchanger. The concentrated solution (three percent solids) is sent to an incinerator. At the end of a batch, the recirculation pump is used to transfer the concentrated solution to the incinerator feed tank. When the process originally started up, the batch evaporation was run every three to four days, whenever enough mother liquor had accumulated.

The process worked very well. Recently, due to increased production, the solvent recovery runs almost every day and the heat exchanger plugs up before the batch is finished. However, if the operators catch the fouling in time and let the solids settle, the run often can be extended. The solids content exceeds 10 percent, though, and the solids settle quickly, in under 15 minutes in a graduated cylinder. The incinerator can handle the higher solids content but only has limited spare capacity. Also, a boost in production is anticipated, and the solvent recovery may need to be nearly continuous. We know that increasing the amount of concentrate to the incinerator is not desirable due to solvent losses and environmental regulations. Can you suggest a solution?

Recycle Concentrated Solution

Having a solvent recovery system on a crystallizer is an ideal situation to recover the solvent and the solids from the mother liquor. Based on the problem description, it would be correct to assume that the solids in the solvent are the same as the solids that are being crystallized. If this were the case, my first choice would be to recycle the concentrated-solid-containing solution to the crystallizer. The temperature and flow rate of the concentrate would have to be controlled such that it does not disturb the crystallization process. If the crystallization process is run continuously, recycle the concentrate as crystallizer feed. You will gain two benefits, total recovery of the solvent and the solids, an ideal solution.

Making Sense of Temperature Sensing

Problem

Why did a thermowell in a recirculation pipeline near the discharge of a pump indicate a change in temperature sooner than a thermowell inserted via a top nozzle about halfway down into a well-mixed vessel?

Poor Mixing

This can be explained in the following ways: the thermowell in the tank may not be in contact with the liquid or the contact is such that the liquid in not representative of the mix—it may be sitting in the tank vortex; the tank may not have baffles, thus mixing is not at optimum; two different temperature liquids are being blended but do not have sufficient residence time in the tank to be homogeneous; or the liquid at the discharge of the tank has gone through two very turbulent zones—these include a vortex and the mixing of liquids in the propeller.

Mixing has produced a temperature that is different from the probe in the tank by blending the two different temperature liquids. It is possible that the final equilibrated temperature of the tank is different from the temperature read by the thermometer at the discharge of the pump.

Boost the Reliability of a Solvent Supply Pump

PROBLEM

Toluene is used in manufacturing paint. Fresh solvent is pumped from a tank farm (Figure 2.3) to the color tanks. Tanks 1, 2, and 5 perform without a problem, but tanks 3 and 4 have a problem with both pumps 1 and 2. Cavitation is especially bad during the summer months; the

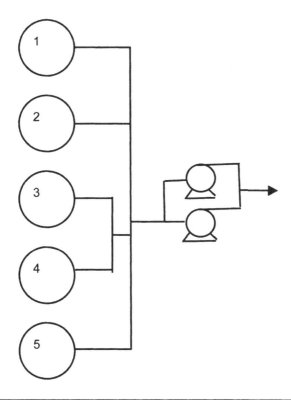

Figure 2.3 Pump cavitation problem.

tanks and pumps are outside. Pump 2 is the most troublesome. Seals are replaced twice as often on pump 2 as pump 1. The pump inlet flange is directly connected to the concentric reducer for pump 2, but there is a 10-inch-diameter spacer between the reducer and inlet flange for pump 1. Both pumps were installed with concentric reducers instead of eccentric reducers to save delivery time for fittings. Any suggestions on what can be done to improve the operation?

Use Eccentric Reducers

This cavitation problem is occurring due to a vapor pocket that is created in the reducer. Summer heat is vaporizing enough toluene to accumulate in the concentric reducer and the 10-inch-diameter spacer. This vapor is being sucked into the pump, thus lowering the NPSHA (net positive suction head available). There are two potential ways the problem could be solved. They could be used in combination or individually: One is to replace concentric reducers with eccentric reducers. If concentric reducers are the only choice, then vent the vapor from the problem fittings to the storage-tank vent. This will relieve all generated vapor to a safe place. If the reducers are replaced, install the eccentric reducers with the flat side on top.

Uncovering the Cause of Solid Buildup

Problem

A vibrating screen separates pulverized lumps of coal from dust (Figure 2.4). The screen is fed from a lock-hopper above. The fine particles must be separated because they could become carryover for a gasifier (entrained-flow); a pelletizer scheme is being tested to reuse this material. The lumps are then fed to a screw conveyor to the gasifier. On a vibrating screen, solid is supposed to be spread evenly across a mesh screen. The rocking motion and the slight slope cause the solid to flow towards the discharge and into either the pneumatic conveyor pickup pipe or the hopper for the screw conveyor. This isn't happening. Instead, coal is building up on the screen and blocking flow from the lock-hopper above. The screw conveyor hopper is plugged and the pickup pipe is plugged. Adjustments to the slope and the rocking motion have been increased to no avail. Any suggestions?

Look at the Material Balance

Based on the problem description the system is under-designed. A combination of the following is a possible cause of the problem: the coal feed

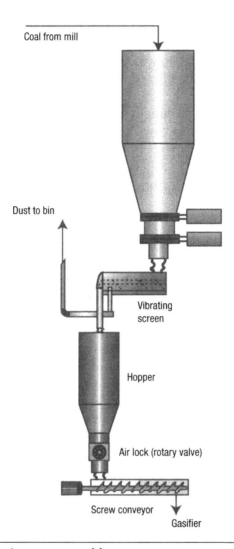

Coal from mill

Dust to bin

Vibrating
screen

Hopper

Air lock (rotary valve)

Screw conveyor

Gasifier

Figure 2.4 Vibrating screen problem.

rate is too high for the size of vibrating screen; as the coal moves, the further size reduction is generating additional dust; not enough air is being used to remove dust from the system; air impingement nozzles aren't properly placed; and dust bridging is leading to plugging in the hopper.

Coal buildup on the screen may suggest that there is insufficient slope for the coal to move forward.

If the figure is correct, the orientation of the dust intake may be wrong. Pickup pipe should be straight up so that all the dust can be directed to the dust collector without any accumulation or bridging in the pipe. Air has to impinge at a high enough velocity to dislodge the dust and carry it to the pickup pipe. In addition, it might be necessary to have air flow countercurrent to the coal flow along the vibrating screen bed so that all the dust goes to the dust collector with a minimum of dust entering the hopper.

The system should be redesigned with a good mass balance. I believe a proper design with above considerations taken into account can solve the problem.

Consider Options for Automation

PROBLEM

We operate an acid regeneration plant (ARP) for a steel mill. In the process, $FeCl_2$ solution is oxidized to Fe_2O_3 and HCl in a spray roaster; the HCl is recycled to the mill, and the oxide is agglomerated and pelletized for feeding back to the mill (Figure 2.5). Currently, operators do laboratory tests to monitor the quality of HCl (target 18.5 percent). The absorber hits an azeotrope at 20.4 percent, at which time the scrubbers become absorbers and we will be fined for an acid emission. Someone suggests monitoring the concentration of the acid in the absorber, e.g., by inferring concentra-

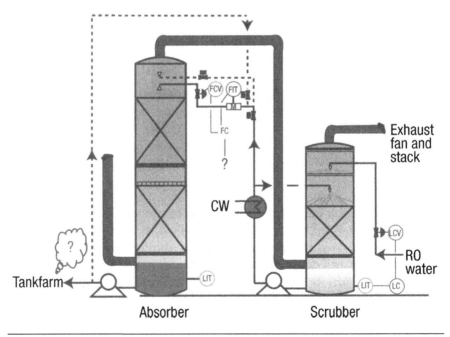

Figure 2.5   Automate make-up to absorber.

tion by measuring conductivity or density or measuring directly by using an automatic titrator. Automating the absorber is also appealing because this loop requires constant attention, especially during startup and shutdown. What should we do? Keep in mind that ARPs are water-restricted.

Automation can also Boost Morale

For efficient and economic operation of the ARP, control of the acid concentration is necessary. Thus, I would recommend that the possible control technology choices that you have outlined and others that are available on the market should be compared, and the best operational and economic option should be selected. Savings during startup and shutdown operation, along with the need for constant attention, will justify the expense. In addition, inclusion of an automatic control scheme would be a morale booster for the operational people. This has benefits that are difficult to dollarize in the justification.

Sticky Valve

Problem

During our plant shutdown a flow control valve was replaced. The globe valve charges a reactor used for making a volatile organic chemical, an intermediary to a polymer (Figure 2.6). The previous valve faithfully served for several years, though we had trouble with it once because the heat tracing failed. The board operator said that flow dropped to about half, which took longer to fill the reactor. Although the new control valve served well for a few weeks, it's now sticking—and it's getting worse. For some reason, the flow problem also has returned. Obviously, we need to pull the valve. What should we be looking for?

Focus on Heat-Tracing Failure

From the problem description, it seems that some of the organic material fed to the reactor or being produced in the reactor is vaporizing and might be sticking and building up on the outlet of the valve, thus blocking the flow.

The heat tracing used on the valve might not be adequate to melt the material stuck at the outlet of the feed valve. If you remove the valve and see a material buildup, it will confirm a lack of heat to keep the material fluid. The fact that the old globe valve had a problem due to tracing failure also suggests that failed heat tracing led to feed problems. It also confirms that the organic material freezes and slows the flow.

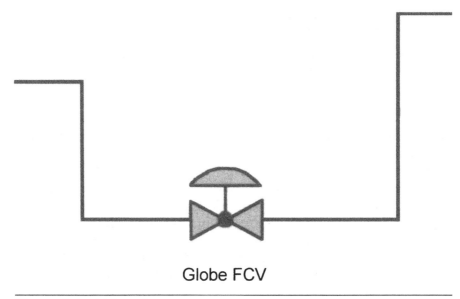

Globe FCV

Figure 2.6 Sticky valve.

Since the product isn't being affected, it's a confirmation that all of the necessary feed material is going in but taking longer due to blockage. Self-regulating heat tracing, where the temperature can be controlled above the melt point of organic material, should alleviate the problem.

Of course, it's possible that the problem is engineering. One would expect that the engineer should use the valve with similar attributes, which had worked earlier. The valve should be removed and inspected. If the fault lies with the vendor or the engineering, it will be quickly realized after disassembly.

Quick Fix Needed!

Problem

A decanter serves a distillation column used to separate water from solvents. Recovered solvents are then burned in a thermal oxidizer (TOX). The decanter separates the heavy organics, such as carbon tetrachloride, from water. The tank diameter is 48 inches and has a volume of 1000 gallons. During startup, it is noticed that the minute the pump begins to fill the tank, the high level switch, 90 degrees from the tank inlet, trips the pump off (Figure 2.7). What can be done quickly to eliminate this problem and keep the overflow safeguard?

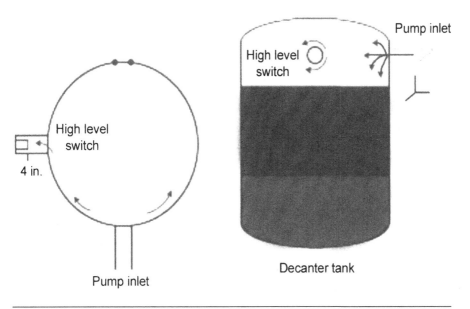

Figure 2.7   Decanter redesign.

EVENTUALLY, DO A REDESIGN

The problem description suggests that the liquid level in the tank is too high when the liquid isn't being decanted. The design has flaws. I would consider the following:

1. Lower the liquid level when the feed pump starts so that there's no splashing of the liquid to activate the high level shutoff.

2. Install a 90-degree elbow to introduce the liquid at a good distance below the high level sensor. This will prevent any splashing contact.

For the long term, I would modify the design. Liquids with the density difference of water and carbon tetrachloride and similar liquids can be separated by using a natural gravity decanter with a proper design. It works extremely well and does not require such a high-volume tank. Due to the high-density difference, the separation is usually quick and clean. Such decanters work well without any problems automatically, and continuously. No level controllers are needed; I have designed many decanters and so can state with experience that this one needs to be redesigned.

Clarify a Clogged Caustic Conundrum

PROBLEM

We have 5 percent caustic solution as a feedstock for a saponification reactor. To prevent freezing, the one-inch-diameter pipe from the storage tank to the reactor pump is heat-traced and insulated. Nevertheless, this pipe frequently is plugged with caustic. What is causing this and what can we do to reduce the maintenance of this line?

TRY RECIRCULATION

The temperature of the solution is not defined in the question. If the solution is not flowing for a while, it is possible that the solution has heated up, causing it to precipitate. Over time this could cause plugging. There are two solutions:

- Reduce the heat flux and inspect the insulation to prevent any caustic precipitation.
- Recirculate the caustic solution to the feedstock tank by putting a T after the pump with appropriate valves so that there is constant circulation of the caustic solution.
- Compare the economics of the two solutions. Also check the physical properties of the caustic solution to make sure there is not salting out.

Succeed through Thick and Thin

PROBLEM

A viscous ingredient is pumped through a heat exchanger to raise it to a sufficient temperature so that a reaction can occur when it then goes to a batch reactor. Unfortunately, frequent nuisance alarms and trips are associated with this system. Typically, startup requires the operator's full attention with all controls in manual mode. Can you suggest some ways to reduce or eliminate these nuisances while assuring a heated exit stream?

MAINTAIN THE TEMPERATURE

The simplest solution would be to heat and maintain the liquid at the desired temperature. The liquid can circulate to the raw material tank. When it is time to charge the reactor, use the existing system, which can be simplified by removing some of the controls that now would become redundant. Maintaining the temperature should allow automatic control.

Quell a Quality-Control Confrontation

PROBLEM

A batch process that consists of a series of feed tanks for a fluidized rotary dryer relies on pH probes to control product quality. There are two steps: pH adjustment, then chemical addition. The pH is adjusted from 1.0 to about 6.5. If the pH is too low, the acid bound to the solid will corrode the dryer and the customer's equipment. Also, low or high pH will cause a problem with another ingredient added to the customer's product. Unfortunately, we can't depend on the probes. They suffer from slow response, especially with negative drops—for example, from 1.0 to 1.1. Also, because of poor residence time in the process, some chemicals added during the pH adjustment have caused problems such as silicon gelling of the probes (in acid solution). The quality-assurance department is convinced that we should be able to meet our customer's needs if we can achieve an accuracy of 0.25 pH—i.e., both probes agreeing that closely. The pH probes last about a week with some product runs and only a few hours with others. Can you suggest improvements in this process or how it's controlled?

IMPROVE THE DESIGN

I believe the process design needs a review. Addition of solid caustic in three different tanks suggests too many variables, and the process will be out of control, as is being experienced.

My solution assumes that chemicals are miscible liquids and will be added only after the desired pH has been achieved.

Using the flow diagram in Figure 2.8 and not having much information about the liquid concentration of the slurry, I would make the following changes to have better process control:

- Use liquid 50 percent caustic instead of solid caustic. If solid caustic is not completely dissolved due to uneven mixing, variable pH readings will result; this is due to incomplete and delayed dissolution of solid caustic.
- Use only one tank with a pump-around system and sufficient residence time for proper mixing; this should give a stable reading. The pH probe would be inline and controlling flow of liquid caustic into the tank. It will have to be strategically located.

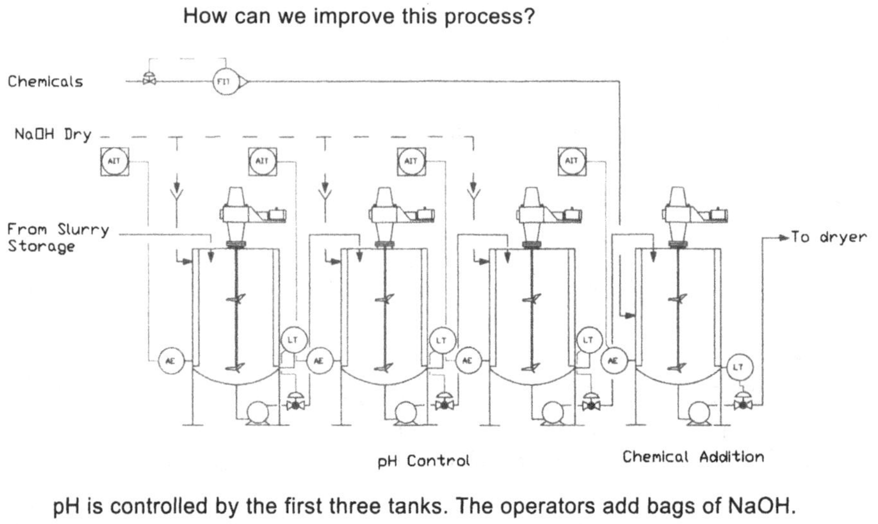

How can we improve this process?

Chemicals

NaOH Dry

From Slurry Storage

To dryer

pH Control   Chemical Addition

pH is controlled by the first three tanks. The operators add bags of NaOH.

Figure 2.8 Why is it so hard to meet quality requirements?

Get out of the Soup

PROBLEM

We add a polymer agent as a coagulant to a cosmetic tank. During development, the ingredient was added by hand. In a large bulk tank, this is too inconvenient. We enlisted an open-impeller centrifugal pump to batch the agent to the bulk mixing tank—i.e., reactor. Now, however, the product, which should have a consistency of tapioca pudding, resembles runny soup. How should we investigate this problem and what could be the cause? How do you suggest we can improve the consistency?

USE AN EDUCTOR

It seems that process development work has been done well but things are not working as planned on the commercial scale. The problem as stated does not tell how and where the coagulant is being added and whether it is a liquid or a solid. If it is a liquid, it could be added using an eductor on the inlet side of the pump. This should disperse the liquid coagulant as it goes through the pump. If the coagulant is a solid, it can be added using similar devices that are commercially available. They generate sufficient vacuum to suck a liquid or solid. In both cases, the addition should be close to the cavity of the pumping device.

KAYO a Cascade Control Complication

PROBLEM

We react two organic chemicals in a stirred tank. The reaction is exothermic and highly sensitive to temperature. We control the temperature by adjusting the feed flows, particularly reactant A, which makes up 75 percent of the flow. Reactant B is ratioed off reactant A. Our new control engineer thinks we need to program some lag in the control valve for A. His first idea was to install an electric valve positioner on control valve A. The product quality has declined. Is he right about the lag? What other improvements should we consider?

REDESIGN THE PROCESS

Since no information was given on the volume of the tank and flow rates or whether this is batch- or continuous-process, it is assumed that this is a continuous process and the flow is controlled based on the level transmitter. It is also assumed that the reaction is zero order and the reactor has sufficient residence time for reaction completion.

Based on the problem description and the diagram presented, the jacketed reactor with its mixing is not able to keep the reactor at or below the desired temperature. If the reaction is zero order, it will be complete very quickly with its exotherm. This exotherm can easily be contained by having a heat exchanger in the line going to the tank following the pump. Both reactants would be added in the line just before the heat exchanger. Exotherm can be used to accelerate the zero order reaction, and the heat exchanger can be used to control the temperature. Outflow from the heat exchanger can go to the tank for any further cooling if that is required. If no additional cooling is required, a partial stream from the heat exchanger could be recirculated to the inlet of the pump, and the reactor might not be necessary. This will convert the exiting batch/continuous process to a highly productive continuous process.

An alternative is shown in the schematic diagram (Figure 2.9). The pipe length could be sufficient to provide the necessary residence time. This reactor could be eliminated, as it is just a wide spot in the line.

Use of Azeotrope to Improve a Process

PROBLEM

A chemical intermediate was prepared by reacting chemical "A" in presence of a catalyst with formaldehyde. Chemical "A" was also being used

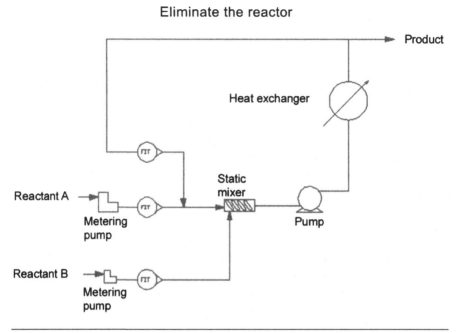

Figure 2.9 Alternative continuous process.

as a solvent for the reaction. Mole ratio of chemical "A" to formaldehyde was 14. Reaction yield was about 50–55 percent. A considerable amount of byproducts were produced and ended up as distillation bottom (waste) to be discarded to a landfill. Using a 1600 liter reactor, reaction produced was about 38 kilos per six hours. Reaction was carried out about 70°C. An overhead condenser was used to reflux chemical "A." Figure 2.10 shows a simplified block flow diagram.

ALTERNATE PROCESS

The catalyst and formaldehyde were miscible but did not react with each other. This property was capitalized to finesse the reaction resulting in higher productivity and yield for the reaction. Chemical "A" was added to the reactor and brought to boil. Formaldehyde and the catalyst mix were sprayed in the vapor space. An artificial excess of chemical "A" was created in the vapor space by boiling chemical "A" and it reacted to produce product "Y." Product "Y" and chemical "A" formed an azeotrop at the reaction temperature. They were condensed. Product "Y" and chemical "A" were immiscible. They were separated using a decanter. Product

"Y" was removed and chemical "A" was recycled back to the reactor. In this alternate process, the yield improved to about 85+ percent, mole ratio chemical "A" to formaldehyde dropped to about 1.5 and the reaction produced about 25 kilos of product per hour using a 250 liter tank. A schematic of the alternate process is shown in Figure 2.11. Along with yield and productivity improvement, significant waste reduction was achieved.

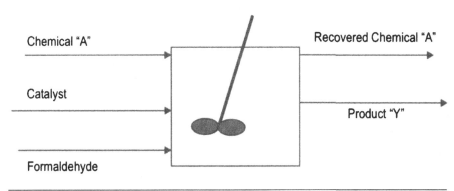

Figure 2.10   Block flow diagram of existing process.

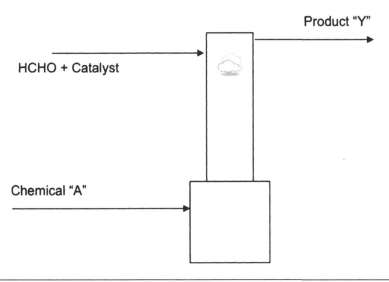

Figure 2.11   Alternate process for product "Y."

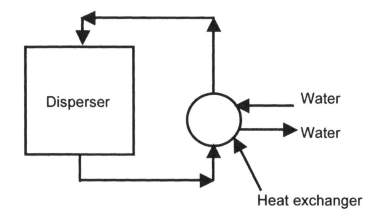

Figure 2.12   Use of an external heat exchanger with coatings disperser.

USE OF HEAT EXCHANGER IN THE MANUFACTURE OF COATINGS

In the manufacture of liquid blends, cooling of batch during the grinding operation is not routine. However, with stringent environmental laws, the need to control the temperature during the grinding to reduce evaporation of volatile organics is becoming an increasing necessity. This is the simplest way to use an inline heat exchanger that will control the batch temperature to below a prescribed temperature, and complete the grinding in minimum time.

Inclusion of a heat exchanger in the grinding step is illustrated in Figure 2.12. The single pass process of flow liquid prevents any settling of solids and minimizes any organic emissions. A minimum amount of solvent is needed for cleaning and can be used in the process.

REFERENCES

1. Malhotra, Girish. "Batch or a Continuous Process: A Choice." *Pharmaceutical Processing*. March 2005;16.
2. Malhotra, Girish. "Less Is More in API Process Development." *Pharmaceutical Manufacturing*. July/August 2005;50–51.
3. Malhotra, Girish. "API Manufacture Simplification and PAT." *Pharmaceutical Processing*. November 2005;24–27.
4. Malhotra, Girish. "QBD: Myth or Reality?" *Pharmaceutical Processing*. February 2007;10–16.

5. Malhotra, Girish. "Continuous Processes Maintain Profitability." *Drug Discovery and Development.* June 2007;30–31.

6. Malhotra, Girish. "Big Pharma: Who's Your Role Model, Toyota or Edsel?" *Pharmaceutical Manufacturing.* June 2007;40.

7. Malhotra, Girish. "Personal Dose: A New Direction and Profit." *Drug Discovery and Development.* December 1, 2007.

8. Malhotra, Girish. "Implementing QbD: A Step-by-Step Approach," *Pharmaceutical Processing.* February 2008;16–18.

9. Malhotra, Girish. "Pharmaceutical Manufacturing: Is It the antithesis of Creative Destruction?" *Pharmaceutical Manufacturing.* July 2008.

10. Malhotra, Girish. "Pharma Convergence: Challenges in Drug Development and Manufacturing Methods." CHE Manager Europe 10/2008. September 18, 2008;26.

11. Malhotra, Girish. "API Manufacturing: A Road Map for Green Chemistry and Processes." pharmaQbD.com. October 14, 2008.

12. Malhotra, Girish. "APIs Simplicity by Design. A Low-tech Approach for Quality and Profitability." *Mastering Process Chemistry.* November 17–19, 2008.

13. Malhotra, Girish. "Alphabet Shuffle: Moving From QbA to QbD—An Example of Continuous Processing." *Pharmaceutical Processing.* February 2009;12–13.

14. Malhotra, Girish. "Rx for Pharma." *Chemical Engineering Progress.* March 2009;Volume 105;No. 3;34–38.

15. Malhotra, Girish. "Hesitation In the Drive to a Continuous Pharmaceutical Manufacturing Process: Real or Imaginary?" *Pharmaceutical Processing.* July 2009;12–15.

16. Malhotra, Girish. "Pharmaceuticals, Their Manufacturing Methods, Ecotoxicology, and Human Life Relationship." *Pharmaceutical Processing.* November 2007;18–23.

17. Malhotra, Girish. "A Fine Chemical Version of Chernobyl? Patancheru, India: An Opportunity for Quality by Design and Environmental Sustainability." http://pharmachemicalscoatings.blogspot.com/2009/02/patancheru-india-opportunity-for.html. Accessed April 22, 2009.

18. "IMPACT: Tons of Released Drugs Taint U.S. Water." http://www.forbes.com/feeds/ap/2009/04/19/ap6308793.html. Accessed April 22, 2009.

19. *Chemical Processing.* 2006 to 2009.

Chapter 3

Commonalities of Businesses

Chemicals are used in many ways and in many applications. Nonetheless, I would like to recognize the number of commonalities that exist. Use of chemicals can be broadly classified in the following categories:

- Chemicals that are consumed to enhance life and lifestyle
- Chemicals that are blended and used to enhance lifestyle
- Chemicals that are polymeric materials.

For purists this could be considered an oversimplification, but if we step back and review the big picture, a majority would fall into these classifications. (I have excluded plastic polymers and petrochemicals from the discussion.)

Chemicals in the first two categories have much in common, but they can be differentiated based on how we use them. In the chemical industry chemicals are categorized as fine/specialty chemicals, flavors and fragrances, active pharmaceutical ingredients (API), dyes, and colors and pigments. Others are categorized as papermaking chemicals, electronic chemicals, paints (architectural, industrial, or automotive), cosmetics, adhesives, and food, to name only a few. Chemicals in the third category are polymers/resins that are blended to produce a product. Blending of API and excipients is an application of chemical unit operations that has to conform to strict regulatory guidelines. (Drug formulation methods are excluded from the discussion.)

Fundamentally, most of the specialty/fine chemicals have either taste, smell, disease-curing value, color-imparting properties, food-texture- or stability-enhancing characteristics, or nutritional value. They are all produced either by a chemical reaction or by blending to give a higher use value that could not be achieved otherwise.

Physical behavior of chemicals and process equipment can be finessed to simplify manufacturing and/or blending processes. This might seem to be a difficult challenge, but in reality it is not. Technologies and methods to produce reacted products have significant commonalities, though sometimes we do not recognize them. Similar laws of physics, chemistry, and mathematics apply to the processes—e.g., chemicals and pharmaceuticals are produced by reaction, crystallization, or spray drying. So are food additives (mostly chemicals). Distillation is used to refine petroleum and also in the production of flavors. Pigments are dispersed and blended with other chemicals, polymers, and additives to produce coatings. Extruders are used to produce useful shapes (dry tablets), toys, and even adhesives .

Methods and technologies perfected for one group of products can be used for the manufacture of other products. Imagination and imagineering are helpful and needed for innovation.

PRODUCTS PRODUCED BY REACTIONS

A good example of this method would be an antioxidant such as butylated hydroxy toluene (BHT), which is used in plastics to enhance appearance. BHT is also used to preserve fats and oils in cosmetics and pharmaceuticals. It is also used in jet fuels, rubber, petroleum products, electrical transformer oil, and embalming fluid.

Another example is methyl anthranilate. It is a fine chemical that has a grape flavor and is fit for human consumption. It is also used to produce herbicide, biocide, bird repellent, saccharin, and other specialty chemicals.

Acetyl salicylic acid, commonly known as aspirin, is a chemical that has a disease-curing value. It has other uses also.[1] So do ibuprofen, naproxen, and paracetamol, and the list can go on.

The point of this discussion is that many of these chemicals are produced following methods outlined in textbooks and patents. Chemicals change, but the fundamental preparation chemistries and unit operations are similar whether the chemical is used as a flavor or a color enhancer or an additive or a disease-curing substance. Many can be and are formulated with other chemicals so that they can be easily dispensed.

Since the chemical processes are the same, with only the molecules being used different, we should be able to apply the same synthesis methods independently of the final application. In the preparation and use of these chemicals we have to meet every environmental and safety and health regulatory standard for the product and the process. We can take advantage of common manufacturing processes when we move from our "chemistry-centric" traditions to an "innovation-centric" or "process-centric" path. This might initially look difficult, but it is not. Understanding of the physical properties and reaction mechanisms gives us the ability to create and manipulate sustainable processes that produce quality products.

Examples

Patents, as noted earlier, are an excellent knowledge and information resource. They can be used to reduce the development and commercialization time for new chemicals and products as well as improving manufacturing methods of existing chemicals and products.

Pyridinesulfonyl chloride or benzenesulfonyl chloride compounds are used to produce herbicides, food additives, or pharmaceuticals. A review of the associated patents suggests that the chemistries are very similar. However, though the methods described to produce these chemicals have commonalities, their commercial manufacturing methods will be very different.

The commonality of these compounds is that an organic amine is diazotized and then sulfated in the presence of a chloride of copper in an acidic medium. In U.S. Patent 7,109,203 the diazotization is conducted at 10° centigrade. In U.S. Patent 6,531,605, it is carried out at below 0° centigrade. In the latter the sulfation is done with thionyl chloride at temperatures below 0° centigrade, whereas in the former it is done with sulfur dioxide at 10° centigrade. Sulfation catalysts are different also. Processes described in these patents, if commercialized as described, would be batch processes.

A comparative mass balance (Tables 3.1 and 3.2) shows that some of the raw materials are being used in excess. Unused reactants will have to be neutralized and treated as waste or recovered and recycled. If recovered and recycled, investment will be necessary, and it can be expensive. Thus, the preferred option would be to minimize use of excess materials. Based on the information outlined in both patents, extended reaction times are suggested. It is possible to reduce these times and simplify processes. Some of the simplification methods are discussed in other chapters of the book.

Chemistry of USP '605 commercialized as described would require refrigeration, and the reaction rate would be slower than the chemistry of USP '203. This is due to the fact that reaction in USP '203 is being carried out at much higher temperature. Due to the need of refrigeration, the investment for USP '605 would be higher than USP '203. Both of the chemistries can be safely commercialized using a continuous process at about 30 to 45° centigrade. Table 3.3 illustrates such a chemistry.

Table 3.1 Preparation of 2-chloro-4-bromobenzenesulfonyl.

U.S. patent 7,109,203: Novartis patent: Sulfonamide derivatives used as a pharmaceutical
Reactions carried out at and around 10°C

Diazotization stoichiometry	Sulfation stoichiometry
Amine = 1.0	SO_2 = 25.9
$NaNO_2$ = 0.97	$CuCl_2$ = 0.5
HCl = 14.0	Solvent: Acetic acid
Diazotization at 10°C	

Table 3.2 Preparation of 2-chloropyridine-3-sulfonyl chloride.

U.S. Patent 6,531,605 Zeneca Patent: Pyridinesulfonyl chlorides and benzenesulfonyl chlorides are used in the manufacture of pharmaceuticals or herbicides.
Reactions carried out at -5°–0°C.

Diazotization stoichiometry	Sulfation stoichiometry
Amine = 1.0	$SOCl_2$ = 4.3
$NaNO_2$ = 1.07	CuCl = 0.011
HCl = 11.66	

Table 3.3 Preparation of 2- alkyl sulfonyl chloride.

Another Amine: Used to produce chemicals and intermediates used as food additive and insecticide.	
Reactions carried out at 35-40°C.	
Diazotization stoichiometry Amine = 1.0 $NaNO_2$ = 1.1 HCl = 2.4	Sulfation stoichiometry SO_2 = 0.068 $CuCl_2$ = 0.068

If exactly the same compound (the starting amine is same) were to be used to produce products for three different applications—for example, specialty chemicals, food additives, and pharmaceuticals—the chemists and engineers involved will still be labeled as experts in their respective areas. This is not accurate, since the chemists and/or engineers are producing the same chemical molecules using the same scientific principles.

Equipment costs, methods of production, and selling prices of the products for the above three applications with similar chemistry will be vastly different. By tradition, the cost of the herbicide and the food additive product, if they are commodity chemicals, will be similar and much lower than the API product. This variation would be present even if the general design and materials of construction for the equipment were the same.

Equipment investment and conversion cost for a specialty chemical and/or food additive plant in a developed country would be similar. These plants will operate five to seven days a week with high on-stream time. Investment and conversion costs for a similar chemical that is used as an API would be five to ten times higher if the plant was located in a developed country. The on-stream time of such API plant could be 50 percent or lower. This is due to tradition and dictated by regulatory manufacturing guidelines even if the product quality is the same.

Based on tradition, an API plant most likely would be a batch plant, whereas the plants for the other applications could be continuous-process. In the API plant, the intermediates are isolated and checked for quality after every intermediate step. This is quality by analysis (QBA), whereas the other two specialty chemical plants will be designed to produce quality from the get-go and follow quality by design (QBD) practices. Intermediate sampling can be discontinued if we understand the physical properties, chemistry, and interaction of chemicals involved, resulting in a QBD process. In such plants quality will be checked at the end of the process with minimal in-process checks.

Simplification Strategy

It is possible to simplify chemistries and produce quality products while still following safety, health, and environmental regulations. The above chemistries have the following three commonalities; the difference is how the chemistries are executed:

- Amine reacts with the acid to produce an amine hydrochloride.
- Amine hydrochloride reacts with the nitrite to produce a diazonium hydrochloride.
- Diazonium hydrochloride reacts with sulfur dioxide or thionyl chloride to produce the corresponding sulfonyl chloride.

Amine hydrochloride and diazonium salts are unstable at temperatures above 5°C. However, if they are consumed as soon as they are produced, they do not present any hazard. This is suggested in USP '203. Based on my experiences, the reaction temperatures can be as high as 40 to 45°C if the diazotization process is designed correctly. The yield is excellent.

A strategy of instantaneous use or removal of a formed intermediate chemical is extremely valuable. This and other strategies are discussed elsewhere in the book. If used properly, they can promote and speed up reactions.

It is well known that higher temperatures speed up the rate of reaction. If quick consumption of the formed intermediate and higher reaction rates can be used effectively, it is possible to convert many batch processes to continuous processes. Theoretically, a continuous process will have better yield than a batch process. We all know the advantages and value of continuous processing over batch processing:

- Lower cost
- Consistent quality
- Better business process.

The strategy briefly outlined above and in other chapters can be effectively used for process simplification across the organic synthesis spectrum. It is illustrated for the production of some amines and emulsion polymerization.

Amines and Other Products

BASF[2] presents a reactor design to react phthalic anhydride with ammonia using a tube reactor shaped as a coil to produce phthalimide (see Figure 3.1). The reaction is instantaneous and exothermic. In another scheme using the principles outlined above, when molten phthalic anhydride and liquid ammonia, which is evaporated after appropriate control valve, are combined to meet at a point (e.g., a cup) in a reactor, they will react and phthalimide will be formed. Evaporation of ammonia also absorbs some of the heat of reaction and speeds the reaction. Component stoichiometry can be precisely controlled to produce a quality product. The heat of the reaction can be used to keep the mass molten. Since steam is generated as a reaction byproduct, reaction pressure would have to be controlled. Phthalimide is of high purity, and the melt keeps the mass molten. Liquid phthalimide is removed continuously. The difference in the two processes is that USP '519 uses a long tube for sufficient residence time. In the second process the two raw materials are fed in a dip tube to complete the reaction. The molten product overflows in the reactor and keeps the mass molten, thus conserving energy. The yield is above 99 percent with similar purity. There are many examples of the chemistry being executed in simple and complex ways with the quality remaining the same but the costs significantly different.

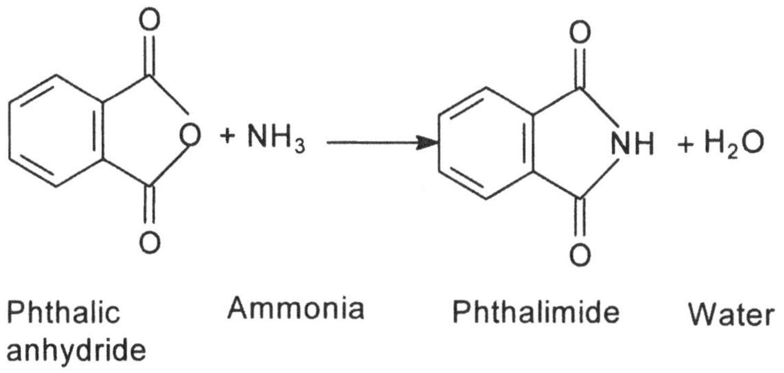

Phthalic anhydride   Ammonia   Phthalimide   Water

Figure 3.1 Phthalimide reaction.

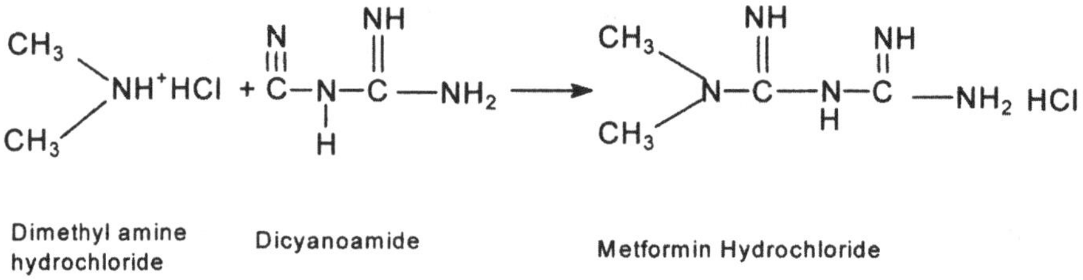

Figure 3.2   Metformin hydrochloride reaction.

Amines from alcohols in presence of a catalyst[3] can be produced continuously using the strategy outlined above to produce phthalimide:

$$RCH_2OH + NH_3 \rightarrow RCH_2NH_2 + H_2O \qquad \text{(Equation 3.1)}$$

Physical properties, kinetics, and residence time will dictate the design of the reactor so that the raw materials can be fed properly and the product removed as soon as it is produced. The chemistry and technology can deliver a green and environmentally sustainable process.

Many chemistries can be single-step processes. The reaction of dimethyl amine hydrochloride with dicyanodiamide is another example that can produce metformin hydrochloride (N,N-dimethylimidodicarbonimidic diamide hydrochloride), a drug used for the treatment of diabetes (see Figure 3.2).[4] Compared to the current methods, this is a high-yield clean process with minimum waste. Both raw materials can be used in molten form. Using stoichiometric quantities with necessary residence time, they will react to produce the desired product. Yield and purity are high. The resulting product is crystallized to produce the desired purity spec product.

Emulsion Polymerization

Acrylic emulsion polymers are the synthetic latexes that are used in paints and other applications. Initiators and surfactants are used in the polymerization. They are produced using a batch process. The commercial processes are an exact duplication of the laboratory process. It is possible to synthesize these latexes continuously. Finesse and the strategies outlined above and in other chapters are needed. One has to understand the physical properties of the chemicals involved and how they react to produce the product.

Generally in the batch process, monomers, the chain transfer agent, and the surfactant are mixed in water and heated to a desired temperature to produce a latex polymer. Monomer addition is controlled to control the polymer formation and particle size.

In 1986–87 I was asked to develop a continuous process for emulsion latex. Skepticism was high, as tradition dictated batch processing. The Smith-Ewart-Harkins theory of free-radical emulsion polymerization can also be used for the development of a continuous process. I used my experience in finessing the physical properties, addition method, and rate of addition of the raw materials to develop a continuous process.

A surfactant and a monomer are two inert components, which do not react with each other but create the dispersion when mixed under agitation. Another solution of initiator in the water is again a mixture of two inert materials. When these two solutions are combined in the desired stoichiometric ratio at the preferred temperature and provided sufficient residence time, they will react and produce the desired polymer. Temperature and mixing have to be controlled for the individual polymer product.

The above scheme manipulates the physical properties, chemical characteristics, and reactivity of the reactants. Since the reaction time—the residence time and temperature—are controlled, the particle size of the polymer can be controlled to a very narrow range. Based on my experiences, this method does produce emulsion polymer continuously with minimum unreacted monomer concentration. Since the conversion is high, the smell of the unreacted chemicals is minimized. This is a cost-effective continuous process producing high-quality product. In addition to being cost-effective, the process requires significantly low investment compared to a traditional plant and is a green and sustainable process. A recent patent[5] confirms the methodology described above.

FORMULATED PRODUCTS

Many chemicals are blended to produce products that facilitate our lifestyle. Excellent examples of blended products would be coatings (architectural, automotive, industrial), cosmetics, paper treatments, inks, fragrances, and so on. As in chemical reactions, one can manipulate physical properties to simplify the manufacturing process; similar finesse can be applied in the production of formulated products. In addition, there are equipment technologies that are commercially available to produce certain types of products that can also be used to produce others and may simplify their manufacturing process. One has to understand the functionality of the equipment to cross-fertilize equipment use.

Method of Production

Blended products are formulated using a batch or continuous process. General business principles would apply in the selection of the appropriate method. It is well known that continuous processes have distinct cost and quality benefits over batch processes if the product volume warrants them.

Surface Coating Manufacturing

Coating (architectural, automotive, wood, industrial, paper) manufacture involves a blending of chemicals. Purists may not like my broad generalization, but many times looking at a glass "half-full vs. half-empty" can bring about innovation. In the manufacture of coatings, chemicals are traditionally mixed in a certain sequence. Solids have to have a certain particle size. I will use architectural coating as an illustration.

We expect architectural coatings for a certain use to be of the same color or shade and to perform in exactly the same way every time. We may buy two different gallons of the same paint on the same or different days. It is very possible that the color and performance of these paints can be different when applied. This can happen due to any of the following reasons:

- Coating is from two different batches.
- Coating is from two different plants of the same company.
- Coatings meet the spec but are from two different ends of the specification, suggesting that the specs are so broad that they can yield two different performance products.

People in the coatings business might not agree with my conjecture. However, my explanation is as follows. Coatings contain many different chemicals, and each supplier has its own specifications range. The paint performance specification is generally set broadly enough to accommodate variation in the raw materials so that when combined they can produce the desired performance coating. This can lead to batch-to-batch variations. Just for these reasons I define batch paint manufacturing as making one can at a time.

I will review the similarities of water-based architectural and paper coatings individually and collectively. Concepts discussed can be applied to other coatings.

How can the problem of color and quality variation, even though minor but discernible, be alleviated? In the case of architectural coatings, off-spec paint can be blended or sold in a "fire sale." There are other potential methods whereby off-spec and waste paint can be used as a raw material. Some were discussed in Chapter 2 and are also discussed later in this chapter. In the case of paper making, if the coating on paper does not meet specifications, paper is heavily discounted or repulped and recycled.

Any recycling or discounting of the product does not deliver the desired profit margins. Producing quality the first time and every time should be the only option. Since architectural and paper coatings are produced in large volumes, consistent quality that meets specifications would deliver higher margins. One of the ways in which this can be achieved is by producing the coatings continuously. I will use architectural coatings as an example of a blended product that can be produced continuously. Concepts discussed here can be extended to other blended products.

ARCHITECTURAL COATINGS

In the manufacture of water-based architectural coatings batch processing has been a tradition. It is three-step process. A pigment grind is prepared using high-speed dispersers. In the grind process water, surfactant, thickeners, dispersant, defoamer, mildewcide, titanium dioxide, and clays are dispersed. After a grind of the desired specifications has been achieved, it is mixed with water and latex, a letdown process. Product specifications are checked; once they are achieved, the coating is shaded with the colorants to produce the desired shade. If the product at an intermediate step does not meet specifications, additions are made based on the experience of the personnel, which can extend batch cycle time. Sometimes the batch cannot be brought in on spec. If this happens, it is either reworked, an arduous process, sold at a loss, or discarded as waste. Disposal as waste is an added expense that can have long-term implications if not done properly.

A variety of clays, surfactants, and thickeners, as well as titanium dioxide, is used to manufacture coatings. This has been the tradition for over 100 years. The process is easy and it works. Paint developers use different ingredients to achieve their product differentiation and respective product performance improvements. Advances have taken place in raw materials in order to meet higher performance standards as well as environmental standards. However, the basic methodology of paint manufacture has not changed much.

Until recently, there were many paint companies in North America and EU countries. These companies would produce niche products to meet the local or regional needs. The batch size would be small. Recent consolidations have progressively led to larger batch sizes to meet market needs. Yet, as stated earlier, the manufacturing methods have not changed.

To meet the seasonal and varying customer needs, paint companies have relied on inventories of finished paints. Since a paint company uses different raw materials to produce the range of paint types, its raw material and finished goods inventory increases similarly. Batch processes result in inventories that are larger than would be needed for a similar business that was producing product continuously.

The recent consolidation of major coating raw-material suppliers, along with that of paint producers, will lead to reduced variation and complexity of both raw materials and finished products. This in turn means that coating specifications and performance are going to be similar. This can present an excellent opportunity to recycle "postconsumer" paint. This consolidation can also facilitate continuous processing.

CONTINUOUS PAINT PROCESSING

Dow product UCAR 300[6] is used to illustrate that it is possible to produce paint continuously. Simplification of raw materials is the first step of the process. Again, to purists this approach might not be acceptable. However, if we have available materials that can deliver different performance coatings, the paint-making process becomes very simple. I demonstrated this in 1985. The number of raw materials needed to produce different performance paints at different price points is reduced and the manufacturing process simplified. This improves the total business process, increases profitability, and improves cash flow, and customers get high and consistent quality coatings for their needs.[7] This method can also work for batch processes.

In its simplest form paint contains water, pigments, clay, latex, mildewcide, dispersants, and rheology modifiers; these ingredients are combined in different ratios to produce different performance paints. This is illustrated in Table 3.4.[6] Formulas in Table 3.4 suggest that there is room for component optimization. As a case in point, each of the paint types uses exactly the same amount of defoamer in the grind as well as the letdown. This could be optimized.

Dow literature suggests use of different pigments and latexes for different performance paints, but for simplicity they have been combined. Coating formulators will understand the value of this combination.

Batch processing could be simplified if there were one grade of TiO_2, one grade of clay, and an additive package that could be blended in different ratios to produce paints with the desired performance. Effort and creativity would be needed on the part of paint formulators to develop such recipes. Due to ingrained traditions, development and rationalization of coatings might take longer than the normal time. This effort would definitely benefit companies that manufacture different performance coatings.

Table 3.4 Formulations for different quality paints.

	Good flat	High Quality Flat	Eggshell	Semi gloss	Deep tone	Ultra Deep Tone
			Pounds			
Grind						
Water	275	275	263	191.6	191.7	225.5
Thickener	1	1		1	3	1
Surfactant	4.5	4.5	2	3	4.5	3
Dispersant	5	5	8.2	5	10	5
Defoamer	2	2	2	2	2	2
Biocide	1.5	1.5	1.5	1.5	1.5	1.5
TiO2	180	200	300	220	40	0
Pigment	347.2	293.6	147	63	362	300
Letdown						
Latex	200	300	395	471.3	280	250
Defoamer	2	2	2	2	2	2
Thickener	12	10.3	5	19.7	10	10
Water	169.8	93.3	51.4	89.2	204.2	200
Total, pounds	1200	1188.2	1177.1	1069.3	1110.9	1000.0

Companies do understand the value of rationalization and simplification. Successful simplification would result in raw material consolidation. It would have a significant impact on raw material suppliers. Specifications of their products would also become narrower. Effort and time must be spent on such an attempt, but traditions have stood in the way of simplifying the coating manufacturing process. Such simplification can easily convert a batch process to a continuous process. Success will improve profitability and simplify the total business process in terms of raw material, in-process, and finished goods inventories. Inventory turns can be as high as can be managed by the operations. Cash flow will improve.

If paint developers cannot move to single grades of pigments, latex, and/or TiO_2 due to product demand, there is another way to produce paint continuously. Again, a formulation development effort would be

Table 3.5 Suggested formulations for extender pigment slurries.

	Minex Slurry	CaCO₃ Slurry
Ingredient	Weight, pounds	Weight, pounds
Water	150	150
Attagel 50	5	5
Silica	25	0
Minex 4	140	0
Clay 1	50	0
Clay 2	0	75
Calcium carbonate	0	125
Dispersant	0.3	0.2
Defoamer	1	1
Total	371.3	356.2

needed. One possible method is to develop different slurries (one for TiO_2, one or two for pigment, and one or two for latex). For simplicity we can call pigment slurries minex slurry and $CaCO_3$ slurry. Table 3.5 suggests potential formulas[8] for extender pigment slurries.

Minimizing the number of raw materials used has another significant benefit. It can minimize waste that is generated by discarding paint (post-consumer, company returns, and off-spec). Minimization of raw materials, together with commonalities in use, can promote the use of surplus and/or waste paint as a raw material for fresh or recycled paint or other applications as discussed in Chapter 2.

Clay slurries, along with TiO_2 and latex, that have been formulated to their own designed specifications can be metered in proper ratios to produce paint continuously. Schematically a simplified process is shown in Figure 3.3. Process stoichiometry can be controlled using commercially available in-line flow meters, ratio controllers, and process control valves. As I indicated earlier, such a process concept yields excellent results.

Another slurry method is to have one common slurry of clay, calcium carbonate, titanium dioxide, and silica at their common level. This illustrated in Table 3.6a-d. As we want to make different performance quality paints, necessary slurries can be metered in to produce the respective coatings. An option of different slurries has an added advantage, as it can allow the addition of specialty latex or additives to deliver any special performance coating.

Table 3.6a shows quantities of different additives that are needed to make five different quality performance coatings. We see some commonalities. Using experimental design methods, we could come up with recipes (listed in Table 3.6b) that could deliver products that have performance characteristics very similar to those from the recipes listed in

Table 3.6 a. Composition of paints; b. Optimized composition of paints; c. Optimized common slurry composition; d. Differential amounts of additives needed to produce different coatings.

Quality level	A	B	C	D	E
Clay	60	80	100	60	60
$CaCO_3$	60	80	100	60	60
TiO_2	100	80	60	120	140
Silica	60	80	100	60	60

a

Quality level	A	B	C	D	E
Clay	60	60	100	60	60
$CaCO_3$	80	80	100	80	80
TiO_2	100	80	60	120	140
Silica	60	60	100	60	60

b

Common slurry	
Clay	60
$CaCO_3$	80
TiO_2	60
Silica	60

c

Quality level	A	B	C	D	E
Clay			40		
$CaCO_3$			20		
TiO_2	40	20		60	80
Silica			40		

d

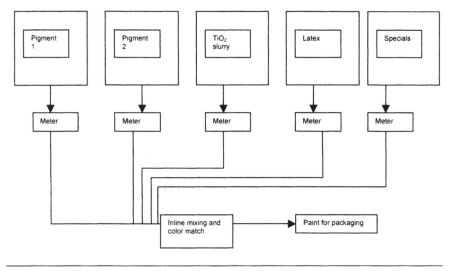

Figure 3.3   Continuous paint process.

Table 3.6a. This would suggest that we could create a slurry, shown in Table 3.6c, that has common elements to produce the five performance level paints (Table 3.6a) and use it, per Table 3.6d, to produce the desired quality coatings. The advantages of these methodologies have been discussed earlier. We need to keep in mind that all liquid raw materials also improve manufacturing productivity and maximize profitability.

Some of the above concepts been patented.[9-11] Product quality can be controlled using in-line wet reflection cells.[12-14] A database correlating wet and dry products would be needed. In-line wet properties and color adjustment methods are feasible and commercial. Since they challenge the prevailing methods, their adoption has been slow and has met with considerable resistance, and an effort to improve manufacturing technologies has to be made.

Cosmetic Ingredients

Lanolin and lanolin oils are used in cosmetics. An example is skin creams. They are a mix of naturally occurring chemicals that are extracted from raw wool that comes from sheep. My reason for mentioning them is to note that the principles of chemical engineering are of benefit and can be applied to extract useful chemicals from non-obvious naturally occurring products.

Lanolin and lanolin oils are extracted from raw sheep wool. Raw wool is washed with specialty surfactants to remove the embedded oils. The water effluent that contains these oils is centrifuged to separate them. Appropriate chemical engineering unit operations are used to treat and purify the separated oils so that they meet the established regulatory standards.

EQUIPMENT COMMONALITIES

Like the similarities among chemical reactions and blending, there are equipment commonalities. Many different types of equipment are used in the manufacture and blending of chemicals. Various equipment types are based on common scientific principles, but with time they are fine-tuned and modified to enhance their operability and performance for the application. It would be beneficial to review these technologies and adopt them to our application. Pumps, reactors, conveyors, heat exchangers, and agitators are only a few of the many equipment types currently in use.

Differences in terminology and measurement are relatively easy to understand and correlate. The challenge comes when technologies being used in two industries are based on the same principles and have been modified for the two applications. Their lineage might be same, but the equipment has refined to deliver optimum performance for each use. Since the developed equipment might look different, it might not be considered suitable for cross-fertilization. This happens when tradition binds the technocrats. It is interesting to review some of the applications that are similar but used in different industries. Following are a few examples.

Tanks, Reactors, and Vats

A simple example will be a vat and a tank. A vessel that is used to hold different mix ingredients for the manufacture of food, cheese, wine, beer, or inks might be called a "vat," whereas the same tank in a paint plant would be called a tank, mixer, or disperser.

The evolution of mix vessels is fascinating in itself. One has to look at the history of products. Since the initial tanks were made from wood, they were called vats. As the size of the vats increased and steel became easily available, vats became tanks. As the fabrication technology and metallurgy advanced, along came the jacketed metal or glass reactors that provide necessary heat or cooling to the processes. If a reaction is taking place in a tank, it is called a reactor. Reactors can be modified to handle pressure or a vacuum. Sometimes, due to the physical properties or corrosivity of the ingredients, special metallurgy or glass might be needed.

We are looking at the development of microreactors. Many patents for different constructions and uses have been filed and granted. However, we have to recognize that they are not the answer to all of the kinetic challenges and will not solve every process problem. Fundamentals of chemistry and chemical engineering (reaction rate, mixing, residence time, heat transfer, corrosivity, etc.) still apply to their use and design. In addition, it is necessary that we compare the economics of their design to what is commercially available. Some of the commercially available equipment technologies can be economically modified and adapted to deliver similar performance. Creative conceptualization is needed, followed up by concept testing. It is a very rewarding experience. However, all too often time to market takes precedence, and expensive processes become a way of life.

Dispersers

In applications whereby solids are to be incorporated in liquids, it is necessary that liquid wet the solids for proper dispersion. Wetting of solids before grinding facilitates the overall process. Coatings, cosmetics, adhesives, and food require that solids be blended in the liquid after a uniform desired size has been achieved. Liquid-solid blending is a mixing unit operation that can produce different performance characteristics if the particle size is not uniform. I have used dispersers as an example to explain the commonality principle.

In the current processes for making coatings, paper, food, cosmetics, pharmaceuticals, and other similar products solids are dispersed in liquids.[15] Dispersers are fundamentally size-reduction equipment whereby a liquid is used as a facilitation medium.

In the standard method of dispersion liquid is charged to the dispersion tank and solids are added. Solid addition is slow, as the disperser performs two functions, wetting and then grinding. Adding solids in large quantities to the liquid vortex can lead to poor wetting and clumping. Gradual wetting with subsequent grinding accelerates the mixing and grinding process. High horsepower machines are used first to wet the solids and then to disperse them to the desired particle size. This has been the traditional method. Since solid addition has to be controlled, it can add to the total dispersion time and extended the batch-cycle time.

Rotor stators are another type of grinder/disperser. They are versatile multifunctional machines that use special configuration blades and stators.[16-21] They are used in processing food, cosmetics, adhesives and sealants, pharmaceuticals, and inks, among other product types. Compared to conventional processes, a rotor stator works for a batch as well as a continuous process. It requires much less energy, time, and investment compared to the standard HSD (high-speed disperser). Small amounts of solids are wet in a small cavity and dispersed. Quick dispersion of solids in a very short time is a deviation from the conventional method. This simplifies the manufacturing processes.

An eductor can be used to facilitate feeding of the solids at a controlled rate and wetting the solids along with a rotor stator mixer or a suitable pump (see Figure 3.4). If appropriate wetting of the solids can be accomplished, a much lower horsepower disperser could be used to finally disperse the solids. This configuration works very well and takes 15 to 25 percent less time compared to the conventional process to produce the desired product. Since the dispersion cavity is small, cleaning, if necessary, is simplified. In addition, a very small quantity of cleaning solvent is needed, which can be reused in the process, thereby making it environmentally friendly.

The use of a rotor stator along with a disperser is not common in coating manufacture, as it is not the tradition. In the evaluation of this and similar technologies, we may find that the initial investment may be higher but can be easily offset by improved throughput. This technique is also ideal for a continuous process, where slurries need to be prepared and preapproved.

Traditionally laboratory equipment has been scaled up to manufacture blended products. It is necessary that various chemical engineering unit operations be applied to the manufacture of blended products.

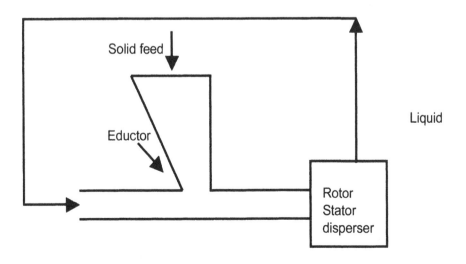

Figure 3.4 Use of inline rotor/stator to blend and disperse solids.

Moving away from high-speed dispersers in existing plants should be considered as companies go forward. Rotor stators will conserve energy, reduce water and solvents needed for cleanup, be environmentally sustainable, and have large impact on the total business process through higher-quality products and improved customer service. A rotor stator, along with an existing conventional HSD, can significantly improve dispersion process and time.

The concepts described above can be extended to other coatings, including solvent-based coatings, inks, cosmetics, paper coatings, and other similar coatings. Continuous review of new technologies and comparison to the technologies traditionally used are a must.

Similarities in Coatings

To purists, tradition is the way to go, as deviations can reduce the performance and efficacy of a coating. This is a concern. However, coatings, raw materials, and their technologies have had a constant evolution. Most of the time developers do not have the time to find the common functions of many of these technologies. Are some of the "performance deliverables" of these raw materials common? If they work well for one application, can they be used in another? We should always look for the commonalities, as there may be significant benefits.

I would like to share an experiment in which I checked similarities and commonalities of water-based paint and paper coatings. I tested concepts for feasibility; they were not subjected to the rigors of application. However, some of the industry data suggests that there are possibilities for validation and commercialization.

USE OF ARCHITECTURAL COATINGS IN PAPER MANUFACTURE

All of us have recycled paper as part of our conservation efforts. During my teens I saw newspapers and printed papers being recycled as paper bags in the Far East. Some paper is still recycled for use as paper bags. Paper pulp without deinking is used as packaging material for fragile goods, e.g., food trays, egg cartons, medical device packaging material, and glass bottle shippers, etc. However, in developed countries, for aesthetic reasons printed paper has to be deinked, pulped, and converted before it can be reused.

My objective in the experiments was to create a process that is environmentally sustainable and simple and that recycles other materials that could impart the desired paper properties. Deinking of paper might be fine for certain uses. However, I wanted to avoid deinking, as I wanted to minimize use of excess water and reduce the water treatment load. If deinking and subsequent processing are done separately, the quantity of the water used increases significantly, and this water has to be treated before the effluent is discharged.

A laboratory disperser was used to pulp the newspaper. Newspaper disperses and pulps well. I added a blend of different shades of water-based interior architectural paint and mixed it thoroughly. The colored pulp (mostly light gray due to residual ink from paper and mixed color paints) was converted to paper using a homemade screen and dried. Even though the paper-making process was primitive, similar to papier-mâché, it did yield an aesthetically pleasing light gray paper. It was smooth and easy to write on. Depending on the shades of the paints mixed, the color of the recycle/waste paint could change. This suggested that there was a reasonable chance of surplus/waste paint being used in the paper recycling process.

Paper coatings use starch as part of the formulation. It would be interesting if the recycle/waste paint, which utilizes different thickeners, can impart paper properties that are similar to those resulting from using starch. Similarly, can any of the thickeners used in the coatings industry be used in paper manufacturing? These options need to be tested. As I have said, it is "out-of-the-box thinking," and it can be of significant value. The use of paint as a raw material might require discussion with the Environmental Protection Agency with regard to waste treatment exemption. Some environmentalists might consider facilities improving the value of surplus paint as waste treatment. However, such a use would be better than landfill, as no one knows the long-term effect of dry paint in soil.

Another way to use surplus or waste paint is as a coating on the paper after it has been formed, such as in lightweight coated (LWC) paper. Table 3.7 shows the comparative composition of blended paints (average of paints in Table 3.4) vs. three different LWC. They are not exactly the same, but a creative chemist/formulator can figure out how to use the surplus paint mix to create a paper topcoat. This can alleviate the challenge of surplus paint that coating manufacturers face. The paper industry could establish a relationship with the paint industry so that the surplus paint can be used in the papermaking process.

If the cross-use of some of the additives used in paper/coating or other industries becomes a reality, it could expand the market for existing products that impart useful properties in other applications.

SURPLUS PAINT IN OTHER APPLICATIONS

Surplus paint offers other opportunities as a raw material. It can be used in the manufacture of drywall and sheetrock. As I have indicated earlier, it might require experimentation and reformulation of the mix.

Table 3.7   Comparison of paint vs. lightweight coatings.

	Avg. paint	Rotogravure	Light weight coated I	Light weight coated II
Water	208.22	338.4	370.0	390.0
Thickener	1.04	1.4		
Surfactant	3.22			
Dispersant	5.78	0.3	0.3	
Defoamer	1.74			
Biocide	1.31			
TiO_2	163.61			
Clay/ Pigment	211.08	564.0	519.6	519.6
Lubricant		5.6	5.2	5.2
Starch			20.8	41.6
Cross linker			1.0	2.1
Latex	286.54	90.2	83.1	41.6
Defoamer	1.74			
Thickener	9.92			
Water	105.80			
	1000.00	1000.0	1000.0	1000.0

Another option for the use of surplus paint is as a dry powder raw material. Surprisingly, water-based paint can be easily spray-dried as a free-flowing solid powder. It is feasible to control the particle size. The solid dry paint can be used as filler in formed (extruded) plastic objects. It could have other uses, which are up to the imagination of the application inventor.

Paint in liquid or solid form can be used in concrete mixing or brick manufacturing.[22] These concepts seem far-fetched, but preliminary experiments show possibilities. Paint has latex polymer, which serves as the binder for the product. Latex polymers can be the binder in cement and brick.

Ink-Jet Printing for Cheaper Solar Cells

Use of ink-jet printing technology[23] to produce solar cells could allow producers to reduce the silicone use by half. Ink-jet printing would reduce the cost of the solar cell by reducing the needed silver in the thinner circuitry. National Renewable Energy Laboratory in Golden, CO has developed the technology.

As I have indicated, imagineering is needed and always welcome.

REFERENCES

1. http://www.fda.gov/fdac/features/1999/aspside.html. Accessed April 23, 2009.

2. USP 4,419,519. BASF. December 6, 1983.

3. C. Gunanathan and D. Milstein, *Angew. Chem.* 2008;120;8789–8792.

4. USP 2,961,377, U. S. Vitamin & Pharmaceutical Corporation. November 22, 1960.

5. USP 6,569,961. Akzo Nobel NV. May 27, 2000.

6. http://www.dow.com/ucarlatex/prod/vinyl/300.htm. Accessed December 3, 2008.

7. Malhotra, Girish. "Continuous vs. Batch Manufacturing," *Paint and Coatings Industry*. February 1999;100–103.

8. Blankshaen, Elizabeth. Private communication. March 9, 2009.

9. USP 4,403,866. E. I. Du Pont de Nemours and Company. September 13, 1983.

10. USP 6,221,145; 7132,479. Coating Management Systems, Inc. April 24, 2001.

11. USP 7,065,429, MicroBlend Technologies, Inc. June 20, 2006.

12. USP 6,533,449. Renner Herrmann S. A. March 18, 2003.

13. USP 6,637,926; 6,288,783. Renner Herrmann S. A. October 28, 2003 and September 11, 2001.

14. USP 7,602,497. BASF Coatings AG. October 13, 2009.

15. Malhotra, Girish. "Dispersion Equipment—An Overview." http://www.specialchem4coatings.com. Accessed May 4, 2009.

16. USP 5,632,596. Charles Ross & Son Company. May 27, 1997.

17. USP 6,000,840. Charles Ross & Son Company. December 14, 1999.

18. USP 6,241,471. Charles Ross & Son Company. June 5, 2001.

19. USP 6,620,234. Millennium Inorganic Chemicals, Inc. September 15, 2003.

20. Langhorn, Ken and Banaszek, Christine. "Charles Ross & Son Company." *Chemical Engineering*. July 2009;40–43.

21. Quadro Ytron. Quadro Engineering Corp.

22 Putting Green Technology Into Bricks, http://online.wsj.com/article/ SB10001424052748704746304574506030258504644.html Accessed November 4, 2009

23. Ink-Jet Printing for Cheaper Solar Cells. http://www.technologyreview.com/energy/22599/?nlid=2009. Accessed May 7, 2009.

Chapter 4

Laboratory Process Development

Development of products that are produced either by chemical blending or by a reaction of chemicals begins in a laboratory. The laboratory development process is just a roadmap. Once a product with a desirable quality and/or performance is developed, it is scaled up and commercialized. It is necessary that we understand the DNA of the various chemicals that are used in the process. It is equally important that we have the knowledge of how they interact with one another to produce the desired product. We can use this knowledge to tweak and finesse the behavior of various chemicals. If we have a complete understanding of their interaction, we can exploit their attributes to our advantage to produce quality product with minimum waste. We can have sustainable processes. The behavioral knowledge of the chemicals can be used in the development of new processes and the simplification of existing processes. Development and simplification begin in the laboratory the day we start to synthesize a new product through reaction and/or blending chemicals.

LABORATORY DEVELOPMENT

Once the reaction chemistry has been identified after a literature search, chemists/chemical engineers test the viability of the reaction, its stoichiometry, and operating conditions based on their experiences in the laboratory. Similarly for chemical blends, chemicals are mixed to achieve the desired performance and cost parameters. The initial parameters are improved upon to the best of the capability of the equipment available in the laboratory. This is the place and time where one has to start looking forward to commercialization of the product. Exploitation of the physical and chemical properties begins here. We have to start thinking about the unit operations and process controls that will be needed in a commercial process. The development process should be such that the isolation of the intermediate products produced by a blending or reactive process is minimized. This can only happen when the product and process developers have a complete understanding of the raw materials, products, and processes.

The following examples illustrate the knowledge building methods. You should not be limited to these examples, which can be expanded to meet individual needs.

Mass Balance

Mass balance is the fundamental knowledge base of every chemical reaction and blending operation. As the development process begins in the lab, it should be prepared. It is a living document even when the product is commercialized. Once the product leaves the laboratory, it is the stoichiometry of the final process that is used by the chemist and/or engineer to commercialize it. The information has multiple uses. I want to share the value of the information here in addition to discussion in other chapters. Heat and mass balance gives a roadmap of the chemist/ chemical engineer's rationale and thinking about the developed product and process.

Mass balance is used in the development and design of unit processes and unit operations for commercial plants. A process technical operating and design manual (knowledge document) can be prepared, which not only specifies the basis but also the how and what of the process. A well-written process technical manual will be used by different functions to manage the operation. Suggested contents of this document are outlined in Chapter 8. A design manual details the design of each equipment type, including piping, pumps, and process control scheme.

I have found that these documents/manuals are an excellent resource for understanding any process and its dynamics. They are an effective training tool and significantly reduce the learning time for chemists and engineers involved with the product. They can also be used for process troubleshooting, improvement, and simplification. A technical manual is a living document. Process operating conditions, parameters, and updates including equipment can be continually refined with the newly generated information.

I was involved in establishing a permit system for existing and new plants for the chemical industry in the Illinois EPA. We suggested that each company should submit a process mass balance for their process and emission-control equipment. Intellectual property was adequately protected. It became a valuable tool for the agency to understand each company's application and operation. It allowed agency personnel to make reasonable and qualitative judgments on permit applications. It actually became an excellent tool for the applicant companies to understand their own operations. Raw materials not converted to a saleable product either were lost as a liquid effluent, emitted to the atmosphere, or sent to a landfill. Reduction of these losses led to process technology innovation and improved profitability for the companies. Companies saw the value and became proactive in conservation.

Mass balance, besides having an economic value, is very helpful in monitoring any manufacturing operation on a daily, weekly, and yearly basis. It is discussed in more detail in Chapter 5.

Chemicals Produced by Reaction

During the laboratory development of a product, it is necessary to test the reaction after each step. Testing is necessary to ensure that the chemical reaction is proceeding as intended and/or to determine the yield of the reaction step. Chemistry developers use different analytical test methods for tracking.

Analytical testing is used to improve the reaction step conversion and the purity of the product. Analytical chemistry and methods allow us to define the chemistry that will produce the highest purity product at the highest yield. Analytical chemistry and the knowledge of physical properties of the chemicals are the fundamental steps in learning the DNA of the synthesis process. Once we have a complete understanding of each reaction step, we should be ready to scale up and commercialize the product.

Development of the right chemistry for the intended product, if done correctly, should not necessitate isolation of the intermediates during commercial production. If we still have to repeatedly test the reaction intermediates and the product after every reaction step of a commercial process, it is an indication that we have not done the scale up correctly and/or do not have control of the manufacturing process for the reaction step.

Repeated analysis of the product produced from each reaction step would be called producing a quality product through analysis or QBA. Anyone who does not have command of the process can still produce a quality product using QBA. However, the process would not be economical. In addition, the QBA method of manufacturing does not incorporate the best manufacturing technology. Actually QBA chokes manufacturing and related innovation (see Figure 4.1).

QBA methodology is the prevailing method of manufacturing a majority of active pharmaceutical ingredients (API) and formulated drugs.

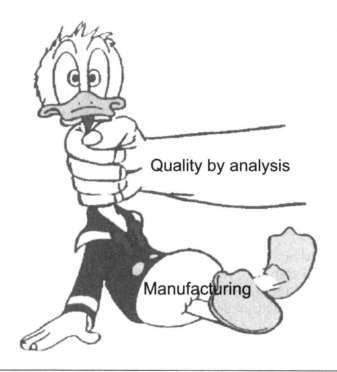

Figure 4.1   QBA chokes manufacturing innovation.

If the product after each reaction step is exactly the same and does not have a significant variation of yield or product quality, such a process would be called quality by design (QBD).

In a QBD process, we will obtain the highest quality and yield of the product at the lowest cost. It will be a sustainable and green process. QBA not only necessitates repeated analytical analysis of the product. It would have variable yield and variable quality, requiring extensive reprocessing and/or remanufacture of the product. QBA processes, compared to QBD processes, have extended batch cycle times, require higher investment, and negatively impact the total business process. Such a process has considerable waste compared to a QBD process. QBD processes can be batch processes, but if we have a complete understanding of the process, they present an excellent opportunity to be a continuous process.

QBD process—a process without intermediate isolation—has another advantage. If an intermediate is toxic, because we are not isolating it, the related investment needed to protect the environment and personnel would not be needed or at least minimized. This can only happen if the process chemistry's mechanism is understood and the process is properly designed. We have to recognize that each chemical can have ill-effects for humans when they are exposed to them in large quantities for prolonged time.

Examples

Value of mass balance is discussed using steps 1 and 2 of U.S. Patents 6,875,893 and 7,057,069. The chemistries outlined in the patents (production of modafinil 2-(benzhydrylsulfinyl) acetamide) suggest a batch process for each step. However, anyone experienced in developing chemical processes can convert this batch process to a continuous process. The value of continuous processing is discussed in Chapter 5.

Patents '893 and '069 present the chemistry of a fine/specialty chemical that just happens to have a disease-curing value and thus is called an active pharmaceutical ingredient (API). The majority of the pharmaceuticals are specialty chemicals (API) that are formulated with approved inert chemicals (excipients) to develop an ethical (brand) and/or a generic drug.

The chemistry as discussed in the referenced patents is straightforward. A careful review suggests that there are process simplification opportunities. How to simplify the various processes was discussed in Chapter 1.

Following are the reaction steps of the process:

Benzhydrol + Thiourea+HBr → S-benzhydrylthiouronium bromide + H_2O

S-benzhydrylthiouronium bromide + KOH → Benzhydrylthiol + urea

Benzhydrylthiol + chloroacetamide → 2-(benzhydrylthiol) acetamide + HCl

2-(benzhydrylthiol) acetamide + Acetic acid + H_2O_2 → 2-(benzhydrylsulfinyl) acetamide (modafinil)

Figure 4.2 suggests the chemistry as outlined in the patent.

Using the information from the patents, a theoretical mass balance is created (Table 4.1) for the first two reaction steps. "Theoretical" (quantitative) means that if everything were perfect we would have 100 percent conversion at each step and the product yield would be 100 percent. The table does not include detailed thermochemical information, but it should be included in a formal document.

Patents '893 and '069 do suggest the use of tetrahydrofuran/water or monochlorobenzene/water mix as alternate solvents. No solvent has been included in the table. Selection of solvent is discussed later, as it is an extremely important part of the process.

Patents outline a chemistry that was practiced in the laboratory to demonstrate the process. Using Table 4.1 as an illustration, a similar mass

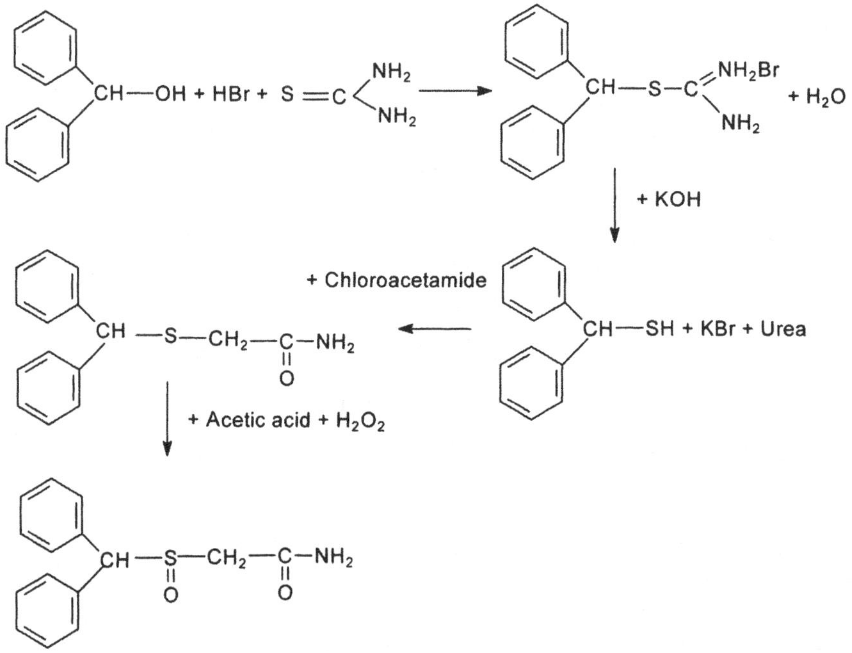

Figure 4.2 Modafinil process chemistry.

balance should be prepared using the chemistry that would be practiced in an actual manufacturing operation.

The stoichiometry listed in the patents is an excellent starting point. Table 4.2 compares the theoretical results vs. the stoichiometry outlined in the patent. These comparative tables do not include any reaction by-products. However, a mass balance for a process should include every starting raw material including solvents, reaction byproducts, and the final product. Mass balance should include physical properties (pH, density, temperature, viscosity, specific heat) and thermochemical properties of each stream, as these properties are used to design the process equipment. Figure 4.3 shows the block flow diagram corresponding to the mass balance in Table 4.1.

We need to make every effort to understand the reaction mechanism and its byproducts. If we understand their formation, we can take steps to minimize the production of undesired products. Such an effort will improve the process yield. This makes the process efficient, reduces waste and its treatment investment, and yields a green process. At times this might not be easy. QBD has to be the modus operandi. Anything short of it is going to increase product cost.

Table 4.1 Mass balance for Steps 1 and 2 of USP 6,875,893/7,057,069.

Stream	1	2	2'	3	4	5	6
Molecular weight	184	81	18	76		56	
Physical State	Solid	Soln.	Soln.	Solid			
Density					1.4	1.51	
Melting point, °C	66			177			
Boiling point, °C	297	126					
Solubility				Water			
Water, gm/L	0.5	Infinite		137			
Viscosity, cp		1.5					
Specific heat, BTU/lb/°F							
Basis: Benzhydrol							
Mole ratio	1.0	1.0	4.875	1		1	
Benzhydrol, lbs	184						
Hydrobromic acid		81					
Thiourea				76			
Water			87.75		105.75	60.7	166.45
S-benzhydrylthiouro-nium bromide					323		
KOH						56	
Benzhydrylthiol							200
Urea							60
Potassium bromide							119
Total	184	168.75		76	428.75	116.7	545.45

Table 4.2 Theoretical and suggested modafinil stoichiometry.

	Theoretical	Suggested
Benzhydrol	1.0	1.0
Thiourea	1.0	1.2
HBr	1.0	1.2
KOH	2.0	2.88
Chloroacetamide	1.0	1.5
Acetic Acid	1.0	3.2
H_2O_2	1.0	2.0

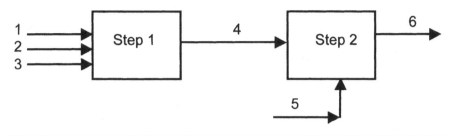

Figure 4.3   Block flow diagram for Table 4.2.

Some of the byproducts from chemical reactions can be toxic. If we can minimize their formation by altering process conditions and/or selecting another synthesis scheme, we can have long-term benefits. Production of higher purity chemical compound becomes easier. This is especially true in the manufacture of APIs. Reduction of toxins also simplifies waste treatment. As the analytical technologies advance, we are able to quantify effects of chemicals on animals, birds, and aquatic life.[1, 2] This will put additional pressure on the chemical and pharmaceutical companies to improve their processes. If processes cannot be improved, greater investment into waste treatment processes will be needed.

In the preparation of a mass balance, a key raw material is used as the basis for the process. Every other chemical is used in proportion to the basis raw material. The main reaction product of each step becomes the basis for reactants for the next step. This is true for every chemical reaction. Mass balance has a similar value for every chemical blending operation. My reason for emphasizing this here is to be proactive in the raw material conservation and optimization process at the development stage. This practice starts in the laboratory; when continued through commercialization, it makes process optimization easy. This practice becomes the knowledge base that is useful in every discussion, product costing, and regulatory compliance issue.

Conservation opportunities become evident only when a mass balance is prepared. As indicated earlier, the laboratory is also the place where stoichiometric optimization starts. We have to recognize that raw materials used in excess not only take up reactor volume but also need to be neutralized, recovered, and/or treated properly before they can enter the waste stream. Any reaction using excessive raw materials requires higher equipment investment for storage tanks, reactors, and associated process handling systems. The product costs go up; profitability can be affected; and competitiveness suffers.

In the illustrated patents every reactant to be added is based on the starting raw material benzhydrol rather than the reaction product of each step (Table 4.2). As a result, the suggested process uses excessive amounts of raw materials. This suggests that there is an opportunity for raw material conservation and improved environmental sustainability.

Considerations for the Development of a Continuous Process

As stated earlier, we need to understand the reaction mechanism and manipulate the physical properties to develop a batch and/or a continuous process. Modafinil chemistry and the physical properties of the raw materials are explored to determine how this chemistry can be converted to a better batch or continuous process.

The first reaction step is the formation of S-benzhydrylthiouronium bromide. Literature[3-7] suggests two possible mechanisms for the formation of the thiouronium bromide. The reaction of benzhydrol with hydrobromic acid to produce bromodiphenyl methane and its subsequent reaction with thiourea to produce S-benzhydrylthiouronium bromide, Figure 4.4, is the most likely route. This mechanism can be used to prop-

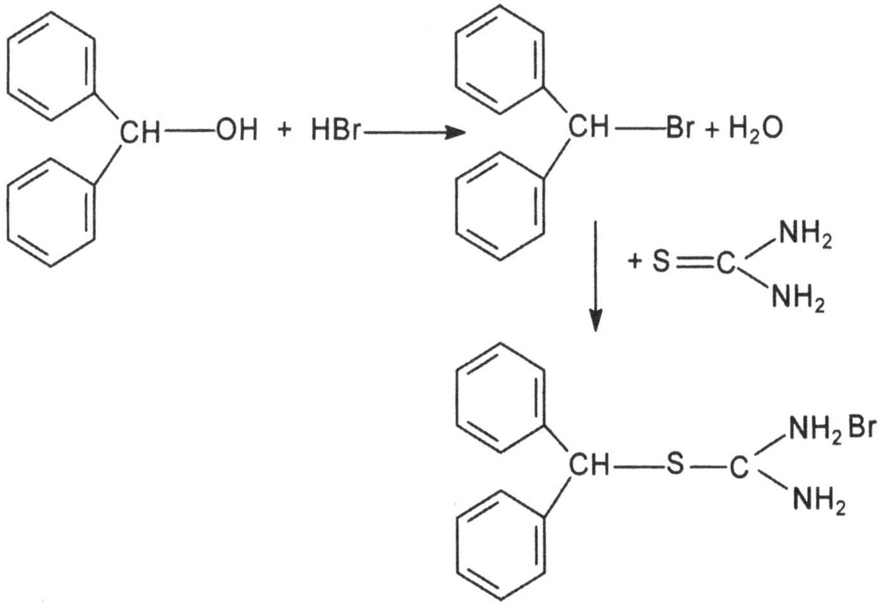

Figure 4.4   Likely route for S-benzhydrylthiouronium bromide.

agate the reaction. Conceptually, we should produce bromodiphenyl methane and control its reaction exotherm to speed the reaction.

Benzhydrol can be fed as a molten liquid, and its flow can be totally controlled. Bromodiphenyl methane is a liquid, and its formation can be used to reduce the system solvent requirement. Bromodiphenyl methane reacts with thiourea to produce the thiouronium bromide salt. This is an exothermic reaction, and the exotherm can be controlled using an appropriate heat exchanger. Thiourea dissolved in water reacts to produce the thiol. Water can be used as a solvent and to slurry the product as a pumpable liquid. A properly designed process might not need any solvent or at the least considerably reduce the solvent need. It is possible that an appropriate solvent might be needed for crystallization. This would make the process environmentally friendly.

Out of the seven reactants used in this reaction, three are solid at room temperature. Benzhydrol can be used as a molten liquid, and the other two are water-soluble. This creates an excellent opportunity to control the process, especially a continuous process. We have to keep in mind that commercial-scale methods are different from laboratory methods, and developers have to envision them and translate accordingly. Figure 4.5 is a possible block flow diagram for a continuous modafinil process.

SOLVENT SELECTION

Solvents facilitate processing. Solvent selection may not seem to be an important exercise. However, it has a significant impact on the overall

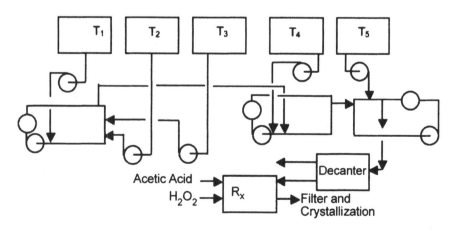

Figure 4.5   Block flow diagram of the modafinil process .

processing of any product that is produced by reacting chemicals or by blending chemicals for an application. Solvent use has to be carefully reviewed. Too dilute a reaction mass will reduce reaction step productivity, and too concentrated a reaction mass will slow the reaction due to reduced mass and heat transfer. It can also create mixing and pumping challenges. If there is too much solvent in a coating, it has to be evaporated. Too much solvent might take longer to dry, or the product might not comply with volatile-organic-compound regulations.

Generally water and one organic solvent are used in the manufacture of specialty/fine chemicals. However, use of multiple solvents, especially in the manufacture of active pharmaceutical ingredients (API), has been a tradition.

For many reasons, such as maximizing process yield, facilitating processing, solvent separation, recovery, reuse, lower environmental emissions, and lower investment and better price due to higher volume, it is advantageous to use a single solvent. Use of single solvent in addition to water might sound like a difficult task. It is a challenge, but the benefits are worth the effort.

The following should be considered in the selection of any solvent/solvents for a process:

- Solubility of raw materials, intermediates, and final product
- Density of organic solvent relative to water
- Solubility of water in the organic solvent and vice versa
- Azeotrope of the organic solvent with water
- Azeotrope temperature
- Specific heat
- Viscosity
- Amount of solvent used

This information in combination[8, 9] can be finessed to achieve an economical process. Processes developed in the lab use excessive amounts of solvent as the process is being developed. There is good reason for this tradition. The laboratory is the place to prove the chemistry and the concept.

Sometimes limitations of the laboratory equipment lead to the use of excessive amount of solvents in a process. Often the tradition lingers in the commercial process, as no one has challenged it. An excellent example of excessive use of water as a solvent is the diazo reaction. Diazotization is a highly exothermic reaction and needs to be controlled. In order to prevent any accidents, inventors have used large quantities of water and ice to maintain the temperature at 0° centigrade or below. This tradition was developed in Europe where dyes were initially produced in the nineteenth century.

In the early development days, jacketed reactors and/or refrigeration systems were either not available or were extremely expensive. With many of the plants producing chemicals located on a river, plenty of water was available. It was economical to use water and ice to control the reaction exotherm. Filtered water, after separation of the solid dyes, would be sent back to the river. In the early days there were minimal water-pollution standards. As time progressed, plants complied with the prevailing environmental laws. However, the use of ice and excessive quantities of water is still practiced even with better technologies available.

Some companies have been granted patents whereby the diazo reaction is being carried out at higher temperatures, suggesting reduced use of reactants[10, 11] and properly designed heat-transfer equipment to control the reaction exotherm. However, the tradition of using excessive amounts of water, ice, dilute acid, and sodium nitrite is still being followed at many companies globally.[12] This is especially true for producing diazo-based dyes. These processes have lower productivity compared to processes whereby the diazotization reaction is being carried out at higher temperatures. With this earlier methodology, we have a significant volume of water that has to be treated before it can be discharged into the water bodies. Solids sludge is filtered and deposited in landfills. The rate of reaction of any synthesis can be significantly enhanced if we can control its reaction exotherm. We just have to understand how we can do that safely.

In the production of colors and dyes, I am told by some of the companies that they have to complete the reactions at low temperatures using an excess of water, as otherwise the product will not have the needed performance. If this is the case, then we need to review and if possible create new color standards so that we reduce the process discharge. Traditional methods of running the diazo reaction and subsequent coupling have to be challenged. However, the tradition overrules alternate methods that are environmentally friendly and improve productivity and profitability. Chemistry prevails. We have to create the habit of conservation; and if we have to forgo some of the pleasantries to pollute less, so be it. Lifestyle changes should be considered.

In the last 50+ years heat exchanger and reactor technologies have advanced. It is possible to process the same chemistry at higher temperatures, thereby reducing plant size, investment, and waste. Higher-temperature reactions, if conducted safely, can result in environmentally sustainable processes with reduced waste.

It is necessary to review how and what can be done to simplify the reaction path and strategy. A recent example illustrates this point. USP 5,958,955 illustrates a chemistry to produce omeprazole at about 76 to 77 percent yield. It uses dry toluene as a solvent. I believe the inventor was not satisfied with the yield. He used water-saturated toluene to carry out the reaction with thionyl chloride. Chemistry tradition is not to use thionyl chloride in a stream containing water, as it will hydrolyze to produce acids. Contrary to the tradition, the inventor used water-saturated toluene and was able to get about 99 percent conversion.[13] This is excellent, as the process can be continuous as well as environmentally sustainable. The inventor used limited mutual solubility of the water in toluene to enhance the mass transfer between phases to improve the conversion. He challenged tradition and achieved better results. This also suggests that we can and should challenge what is considered "tradition" in a safe manner. We might get unexpected and better results.

Many times it is difficult to simulate the conditions of a commercial process in a laboratory. Laboratory equipment provides limited opportunities to manipulate and test alternate process conditions. Due to these limitations, processes using excessive amount of solvents are commercialized. As the processes are scaled up and commercialized, a process of continuous improvement needs to be applied and should result in many improvements. Knowing and using the physical properties of the chemicals can simplify the process. It is necessary to document the why and what of the process—i.e., why certain solvents were selected and under what operating conditions. This assists greatly in the scale-up to commercial process design.

The manufacture of pharmaceuticals has to follow stringent regulatory guidelines. The required protocol is expensive. Reduction of solvent use and for that matter any change in the traditional process method is difficult. It is advisable to incorporate understanding and experience of how to manipulate physical properties of chemicals, process parameters, and conditions early on in the development process. They can result in selection of the economical unit process and unit operation. They can simplify processes and possibly convert a batch process to a continuous process.

If the raw materials/intermediates or the final products dissolve in water or the selected solvent, processing including crystallization becomes easier. Many of the raw materials are solids, thus a properly selected solvent can facilitate their addition to a solution, thereby facilitating the reaction. This can facilitate batch processing and can potentially convert a batch process to a continuous process. Kinetics of the reaction can be influenced, as exemplified in USP 7,227,024 and other patents granted to AstraZeneca. In the laboratory solids are added manually and the reaction proceeds. It is necessary to think and practice the "what and how" of the commercial process. This might be considered cumbersome and tedious during the initial stages of a laboratory development process, but the long-term benefits are worth the effort.

Dissolution of the reaction components changes the density of the organic solvent/water. Their relative density differences are valuable, as they facilitate phase separation. Relative partitioning and the mass transfer of raw materials and products in solvents are of value, as they can be used to accelerate the reaction.

As indicated earlier, the amount of the solvent used in a process has to be carefully considered. If a large amount is used, it reduces process productivity. If the solvent is also a reactant in the process, the reaction productivity can be reduced. This excess can also result in undesirable byproducts, and the overall yield is reduced. An example of such excess is discussed in Chapter 2.

Another consideration in solvent selection is how the various reaction components are going to behave in downstream processing. If the downstream products dissolve in the primary solvent, the reaction kinetics and processing improve.

Since the solvents are recycled, increased dissolved byproducts make processing and solvent recovery a challenge. Sludge from the solvent recovery systems has to be treated and sent to an appropriate landfill. Many times the byproducts of fine/specialty chemicals and coatings, especially APIs, are toxic. Even though the waste is handled according to the prevailing environmental laws, the long-term effect of the reaction byproducts on ecological systems, because of their complexity, has not been studied extensively. Thus it is impossible to postulate and/or predict their prolonged impact.

Analytical technologies have advanced and become sophisticated. Genomic and proteomic approaches to predict toxicity have also advanced, but work is still needed to convincingly predict the long-term effects of these by-products. Some of the ill effects of effluents on the ecology are being seen around Patancheru, India. There might be other similar sites. Thus, it is in the best interest[1,2] of all concerned to maximize yield and minimize the use of multiple solvents.

SOLVENT SEPARATION

If an organic solvent is going to be used along with water, there are two ways to separate the solvent from the mix:

- Phase separation
- Azeotropic distillation.

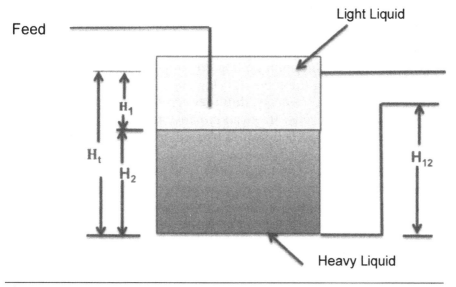

Figure 4.6   Decanter design.

Phase Separation

Use of a gravity separator[14] is the simplest method of solvent and water phase separation. Phase separation is density-dependent, thus it uses no or minimal energy. Since we are relying on the laws of gravity, proper solvent selection does become very important. A centrifugal decanter can be used if the density difference is small and if gravity separation might take too long to separate the two phases.

Figure 4.6 is an illustration of a gravity separator based on density differences. It does not require any automatic level and/or flow controls as long as the solvent density differences are discernable and sufficient residence time is provided for the phase separation. I have been challenged many times on the design and the working of this decanter. The only way to convince is by demonstration, and it certainly makes quick converts.

In the design of the decanter, there is no or minimal pressure drop in the discharge lines. Flow rate and residence time needed for the separation define the decanter volume. Based on the hydraulic balance, the height of the discharge line will determine the liquid-liquid interface. It can be calculated by the following formula:

$$H_2 = (H_{12} - H_t (d_1/d_2)/(1 - d_1/d_2))$$ (Equation 4.1)

Density of the light liquid is d_1 and heavy liquid is d_2.

Azeotropic Distillation

If the reaction product is soluble in either water or the solvent of choice, either could be separated using an azeotrope, preferably a minimum boiling azeotrope. Minimum boiling azeotropes offer an additional advantage of separating the solvents below the boiling point of either solvent. Azeotropes do use energy, thus they add to the cost of the product. However, if used cleverly, they can enhance reaction kinetics, thereby lowering production cycle time for a batch process and facilitating a continuous process.

Another advantage azeotropes offer is in the reaction, especially if the solvent is insoluble in water. This property, along with the phase separation method outlined above, can be used to accelerate the reaction rate. It is commonly used in reactions where the reaction product is water that needs to be removed to accelerate the rate. An example of how azeotrope properties can be used to improve and simplify reaction is dicussed in Chapter 2.

KINETICS

Reaction kinetics is an important part of processes whereby chemicals are reacted to produce a product.

Significant information can be learned while synthesizing the chemical molecules in the laboratory. It is possible to vary reaction conditions and generate yield information to develop a viable and economic process. Experiments suggest process and operating conditions to speed up the reaction. For commercial processes it is of value to have a fast reaction so that the product can be produced quickly and with the highest yield. Some of the methods of improving reaction rates are discussed in Chapter 6.

CHEMICAL BLENDING OPERATION

The majority of blended products contain a mixture of chemicals for the intended performance. Liquid blending is relatively easy compared to blending of solids. Coatings and cleaning-liquid blends comprise the major share of blended products.

Solid-blended products are blend of different solids to produce the desired product. In the solid product category, pharmaceuticals are one of the blended products that have to meet the toughest performance standards. Each tablet has to have exactly the same composition and dissolution rate. In the case of other solid blends (nonregulated), blend consistency depends on the performance requirements. These standards might not be as stringent as the standards for pharmaceuticals.

Liquids that are regulated by government agencies have to comply with tough standards. Architectural coatings have been used as an example for liquid blending. They have to consistently deliver the performance desired by customers. Water-based coatings in general contain varying amounts of thickener, surfactant, dispersant, defoamer, biocide, titanium dioxide, pigments, and latexes. Properly blended, they deliver the desired performance.

If the formulators from the very beginning of the development of liquid coatings think about recycling the leftover coatings as a raw material, this would have a significant impact on the coating business, formulation, and manufacturing technologies. It could create an opportunity for recycling of post-consumer waste. Colored coatings can be used to produce desired colored paints. Solvent-based architectural coating ingredients and formulations could be considered for recycling in similar ways.

Many would consider the concept of coating recycling an arduous task and a risky venture, but it should not be. It is definitely an intellectual challenge that will stimulate "out-of-the-box" thinking. It is also an opportunity to learn about competitive formulation strategies and performance characteristics. From a business and development perspective we could consider post-user coatings as a cost-reduction opportunity that makes available preformulated raw materials at no or minimal cost.

Some of the raw materials that are used for the manufacture of various price-point coatings might have some special characteristics and nuances, but in general they are similar in each category and performance. This is due to the fact that major coating companies are trying to serve the same market. Similarities of raw material behavior and performance should make recycling and reformulation relatively easy. Still, effort would be needed to move away from past traditions.

Recycling of "post consumer" coatings has started[15] by some companies. However, the majority of the coatings are still disposed as a solid to minimize environmental impact. If a government and industry collaborated emission credit program can be established, it would be an environmental and a business win.

REFERENCES

1. Larsson, D. G. Joakim, de Pedro Cecilia, and Paxeus Nicklas. "Effluent from drug manufactures contains extremely high levels of pharmaceuticals." *Journal of Hazardous Materials*. Volume 148. Issue 3. September 30, 2007;751–755.

2. Malhotra, Girish. "Pharmaceuticals, Their Manufacturing Methods, Ecotoxicology, and Human Life Relationship." *Pharmaceutical Processing*. November 2007.

3. Kofod, Helmer. "Furfuryl Mercaptan." *ACTA Chemica Scandinavica*. 7;1953;1302–2306.

4. Johnson T. B., and Sprague, J. M. "A New Method for the Preparation of Alkyl Sulfonyl Chloride." *Journal of American Chemical Society*. 58;1936;1348–1352.

5. Johnson T. B., and Sprague, J. M. "The Preparation of Alkyl Sulfonyl Chlorides from Isothioureas." *Journal of American Chemical Society*. 59;1937;1837–1840.

6. Frank, R. L. and Smith, P. V. J. *Journal of American Chemical Society*. 68;1946; 2103–2104.

7. Private communication with Dr. Charles M. Kausch. June 22, 2009.

8. Gani, Rafiqul, et al., "A Modern Approach to Solvent Selection." *Chemical Engineering*. March 2006;30–43.

9. Wypych, George. "Important Determinants of Solvent Selection." *Chemical Engineering*. June 2006;54–60.

10. USP 7,438,753. Everlight USA, Inc. October 21, 2008.

11. USP 7,109,203. Novartis AG. September 19, 2006.

12. USP 7,439,339. Eastman Kodak. October 21, 2008.

13. USP 7,227,024. AstraZeneca AB. June 5, 2007.

14. McCabe, W. L. and Smith, J. C. *Unit Operations of Chemical Engineering*. McGraw-Hill, Inc.;1956;40.

15. E-coat Kelly Moore Paints.

Chapter 5

Mass and Heat Balance

Mass and heat balance are a critical part of any process, whether it is a batch or a continuous process. Mass balance is an indicator of the process efficiency. For a process whereby chemicals react to produce a new product, mass balance indicates the degree of conversion of the process. For reactive processes it is also the design basis. It suggests what is happening with the process and potential improvements that are possible. For a blending process it suggests how the raw materials can be optimized to produce a low-cost product.

Mass balance has another unique value. Besides its ability to provide the standard factory cost, it can suggest any deviation from it. If a mass balance is done in any commercial plant on a daily basis, it will assist in spotting any leaks, spills, and/or excessive use of materials that can be corrected.

Heat balance is an indicator of the thermodynamics of the process. It is used for energy balance, energy, consumption and design of heat-exchange equipment.

MASS-BALANCE USES

Mass balance has multiple uses:

- It is a visualization of the inventor's thinking and facilitates process understanding.
- It is the design basis of the present process.
- It can be used for process audit and improvement.
- It can be used to improve processing methods.
- It gives the cost basis of raw material for a product.

Table 5.1   Formulas for different quality paints.

	Paint Quality Ingredients, lbs Pigment grind	Good quality flat	High quality flat	Egg-shell	Satin	Semi gloss
	Water	203.0	188.0	263.0	108.3	83.3
Thickener	HEC ER-4400	1.0	1.0	0.0	0.0	0.0
	TKPP	1.5	1.5	2.0	1.0	1.0
Dispersant	Tamol 1124	5.0	5.0	0.0	0.0	0.0
	Tamol 731	0.0	0.0	8.2	5.0	5.0
Surfactant	TRITON CF-10	3.0	3.0	0.0	3.0	3.0
Defoamer	Hi-Mar DFC 19	2.0	0.0	0.0	0.0	0.0
	Colloids 640	0.0	0.0	0.0	0.0	3.0
	Drewplus L-475	0.0	2.0	2.0	2.0	0.0
Mildewcide	Kathon	1.5	1.5	1.5	1.5	1.5
	Na_2CO_3	3.6	3.5	2.0	3.0	4.2
TiO_2	TiO_2 R-900	180.0	200.0	0.0	220.0	0.0
	Tronox CR 828	0.0	0.0	0.0	0.0	240.0
	Kronos 2310	0.0	0.0	300.0	0.0	0.0
Pigments	Huber 70C	223.6	170.1	0.0	0.0	0.0
	Duramite	100.0	100.0	0.0	0.0	0.0
	Nyad 400	20.0	20.0	0.0	0.0	0.0
	Minex 10	0.0	0.0	145.0	0.0	0.0
	Atomite	0.0	0.0	0.0	60.0	0.0
	Polygloss 90	0.0	0.0	0.0	0.0	20.0
	Water	72.0	87.0	0.0	83.3	83.3
	LETDOWN					
Polymer	UCAR 300	200.0	300.0	395.0	334.2	350
	UCAR 6030	0.0	0.0	0.0	137.1	150.0
Defoamer	Hi-Mar DFC 19	2.0	0.0		0.0	0.0
	Drewplus L-475	0.0	2.0	2.0	2.0	0.0
Thickener	Natrosol plus 330	0.0	0.0	5.0	0.0	0.0
	UCAR POLY-PHOBE TR 116	12.0	10.3		2.3	0.8
	UCAR POLY-PHOBE TR 117	0.0	0.0	0.0	17.4	17.4
	Water	169.8	93.3	51.4	89.2	98.0
	Total, lbs	1200.0	1188.2	1177.1	1069.3	1060.5

Mass Balance for Blended Products

Mass balance offers a unique opportunity to understand the mind of the formulator of any blended product. This is especially true when there are different quality products for the same application. An excellent example would be architectural coatings or any other similar product line. Coating formulators design a coating to deliver a desired performance. Once the formula is finalized, it is also the mass balance of that particular product and can be used to manufacture the product.

In the case of architectural coatings and/or similar products, the mass balance has multiple values, especially when different quality performance coatings are produced using similar raw materials. A review of Tables 5.1 and 5.2 suggests that additives from different suppliers are used in the same quantities to products different quality products. Since there are functional similarities, they present the following potential opportunities:

• Raw material consolidation

• Raw material optimization

• Process simplification.

A grouping of functional chemicals (Table 5.2) can lead to potential consolidation of raw materials. If the functional raw materials can be consolidated to a reduced number, preferably to one per performance category, it would be the first step toward simplification of the overall product class. It also becomes an opportunity to implement a simpler batch or continuous manufacturing process. An argument can be presented that, since these additives behave differently, it is not possible to

Table 5.2 Alternate sources of similar raw materials.

Antifoam	Colloid 640 (Rhone-Poulenc) Drewplus L-475 (Drew) Hi-Mar DFC 19 (Hi-Mar Spec.)	Thickener	Natrosol (Aqualon) Cellosize, Polyphobe, HEC (UCAR Emulsion)
Pigments	Duramite, Atomite (ECC) Minex (Unimin) Nyad (NYCO) Polygloss, Huber (Huber)	TiO_2	Kronos R-900 (Du Pont) Tronox (Kerr-McGee)
Dispersant	Tamol (ROH)	Mildewcide	Kathon (ROH)
Surfactant	Triton (ROH)	Polymers	UCAR 300, 6030 (DOW)

consolidate. That might be true, but the opportunity and challenge are to find single or minimal alternates for the same functional chemical so that, when used in varying quantities, they can deliver the desired effect. This is a first step in the simplification of the business.

For the products listed in Table 5.1, opportunities exist to consolidate the dispersants, defoamers, TiO_2, polymers, and pigments. Paint formulators have to exercise creativity to consolidate raw materials.

Once the raw material consolidation has been achieved, the next step would be to optimize the quantity. The design of experiments (1) is a well-known methodology that can be used to optimize raw material quantities as they relate to product performance. Abundant literature is available, and many books have been written on the methodology and philosophy of experiment design. They can assist in the development process and even change the product development methodology at a company.

Consolidation and rationalization of raw materials can also result in simplification of the blending process. Equipment and technologies in addition to high-speed dispersers (HSD) are commercially available that facilitate pigment wetting and dispersion. These technologies expend less energy in the manufacture of coatings—that is, less heat is transferred to the liquids. These technologies reduce emissions especially in the case of solvent-based products. Investment in such technologies is much lower than the conventional technologies that have been used for ages. This reality translates to simplification of the business process. This topic is discussed in Chapter 3.

Current batch processes used in the manufacture of coatings include a disperser tank followed by a letdown tank. Figure 5.1 schematically shows how the paint is produced in the lab and in a commercial plant. In the conventional process raw materials are added sequentially and blended to produce a desired product.

Since consolidation and simplification of raw materials can result in a reduced number of raw materials used, adopting multiple intermediate raw-material mixes, as shown in Figure 5.2, can produce different

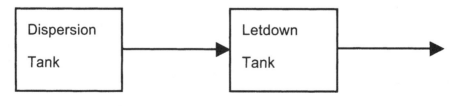

Figure 5.1   Conventional method of making paint.

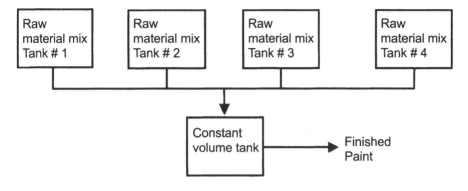

Figure 5.2   All liquid paint making process schematic.

coating formulations. Materials from these tanks can be metered and blended in different ratios to produce the desired quality coatings. Such a manufacturing process can lead to "just in time manufacturing," thereby minimizing raw material and finished goods inventories. This improves cash flow for the business. Raw material consolidation and optimization can also result in purchase-price advantages.

If for some reason complete consolidation is not feasible, a similar concept can be used to add the desired quantities to produce multiple coatings of different performance quality. Such a process would still be simpler than the current practices. Any simplification that can improve a batch process or move the process from a batch to a continuous process will enhance product quality, reduce manufacturing costs, and improve the total business process. These improvements can also reduce waste and create an environmentally friendlier process.

Mass Balance for Reactive Processes

Mass balance for a reactive commercial process illustrates the actual use and production of intermediates and products in any process chemistry. This can be compared with the theoretical mass balance for the same process. Comparison of actual vs. theoretical mass balance suggests opportunities for improvement. Actual mass balance information is used to size process equipment.

Since many of the reactions need solvents, solubility information can be incorporated in the mass balance. This information can be used to select a solvent that can improve the productivity of the process. An alternate place to include solubility information would be in the technical operating manual discussed in Chapter 8.

EXAMPLES

USP 7,227,024

I have reviewed USP 7,227,024 to illustrate the value of mass balance. This patent outlines the synthesis of a benzimidazole compound. The abstract describes it as "a process for the manufacture of omeprazole or esomeprazole from pyrmethyl alcohol via pyrmethyl chloride and pyrmetazole characterized in that the whole reaction sequence is carried out in a solvent without any isolation or purification of intermediates. Further, the reaction is carried out in a solvent system common for the whole reaction sequence and inert to the reactants formed during the process and used in the process and comprises a water immiscible organic solvent and a specified amount of water." The product is commercially sold as Nexium®, Prilosec®, and many other names globally. Production of benzimidazole is a two-step process as illustrated in Figures 5.3 and 5.4.

The first two steps of the patent are described as follows:

- In step 1 pyrmethyl alcohol (8.8 g, 52.6 mmol) was dissolved in toluene (75 ml, water content 0.12 mg/ml) moistened with water (180 µl, 10 mmol) at room temperature. To the stirred solution, at 25 to 30°C, thionyl chloride (8.15 g, 68.5 mmole) was added slowly over 60 minutes (flow rate of 0.083 ml/min). Conversion of the reaction was analyzed with HPLC. Conversion is over 99.5 percent. Water (2.3 ml) was added to quench any excess of thionyl chloride (see Figure 5.3).

- In step 2 an alkaline (13.5 g, 168.3 mmol; 50% w/w sodium hydroxide) aqueous (80ml) solution of metmercazole (9.8 g, 54.2 mmol) was added, followed by additional sodium hydroxide (8.8 g, 110.5 mmol, 50% w/w sodium hydroxide) to reach pH>12. The temperature was allowed to increase to 45°C during the additions. The reaction mixture was left with vigorous stirring for approximately two hours at 45°C. The agitating was interrupted, and the phases were left to separate. The aqueous phase was discarded. The organic phase, comprising pyrmetazole, was washed with water and was analyzed for residues of pyrmethyl alcohol (less than 0.1% mol) (see Figure 5.4).

Table 5.3 compares the theoretical mass balance against the mass balance of example 4 of USP 7,227,024. Comparative analysis shows opportunities. Batch stoichiometry demonstrates the feasibility of the reaction and approximate conditions under which the reaction was carried out.

In the first step, one mole of alcohol should react with one mole of thionyl chloride. However, in the process as described in the patent 1.3

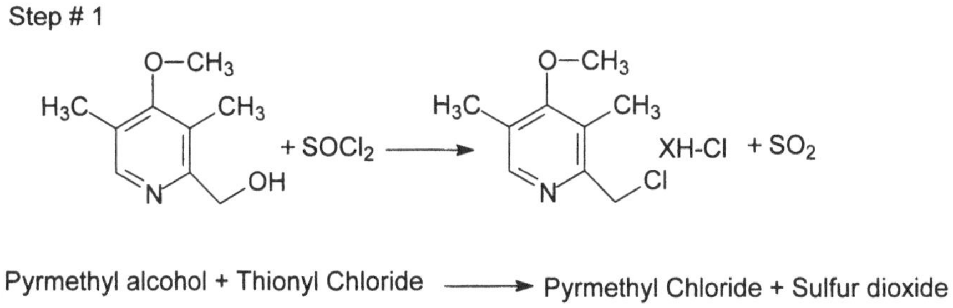

Step # 1

Pyrmethyl alcohol + Thionyl Chloride ⟶ Pyrmethyl Chloride + Sulfur dioxide

Figure 5.3 Step 1 of the Nexium process.

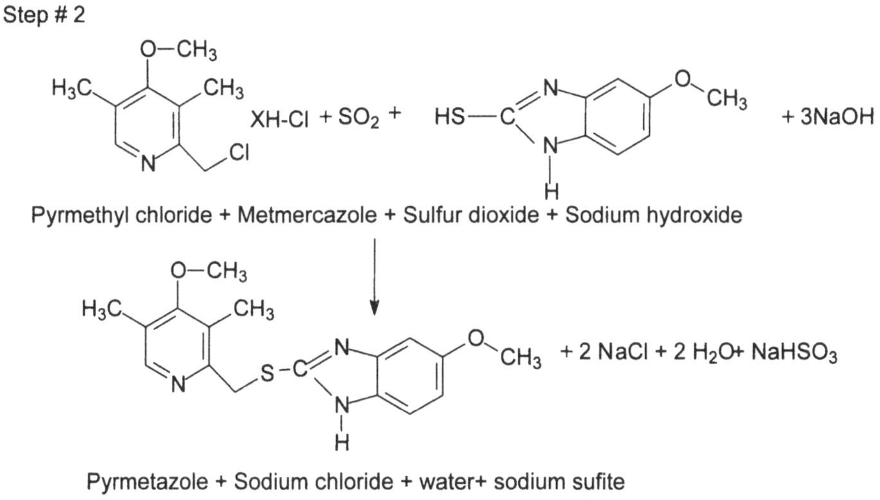

Step # 2

Pyrmethyl chloride + Metmercazole + Sulfur dioxide + Sodium hydroxide

Pyrmetazole + Sodium chloride + water+ sodium sufite

Figure 5.4 Step 2 of the Nexium process.

moles are used. This is a large excess that can be reduced. In addition, the patent suggests dissolving 8.82 grams of pyrmethyl alcohol in 75 ml (density 0.87 gm/ml) of water-wet toluene. However, when water is included in the mix, the alcohol concentration drops to about 3.8 percent. Since pyrmethyl alcohol is readily soluble in toluene, it is very possible to increase its solution concentration for a commercial operation. Higher concentration will reduce the equipment size—lower investment cost and improve productivity. It might be possible to use molten pyrmethyl alcohol as a feed, followed by toluene, to produce a higher concentration reaction mass. This will increase productivity of the process.

Table 5.3 Nexium raw material mole ratios.

Omeprazole				Theoretical	Patent
USP				7,227,024	7,227,024
	Mol. Wt.	Grams	Moles	Mole/key	Mole/key
Pyrmethyl alcohol	167.00	8.82	0.0528	1.0	1.00
Thionyl Chloride	119.00	8.15	0.0685	1.0	1.30
Toluene	92.00				
Permethyl Chloride	222.00				
Metmercazole	180.00	9.80	0.0544	1.0	1.03
NaOH	40.00	11.15	0.2788	3.0	5.28

The patent suggests a certain feed rate for thionyl chloride. The reaction is carried out at 25 to 30°C. Toluene and water mixture azeotropes at 85°C, and thionyl chloride boils at 76°C. The azeotrope temperature and thionyl chloride boiling point are higher than the suggested reaction temperature. Thus, it is very feasible to raise and control the reaction temperature. We have an opportunity to take advantage of the physical properties of the reactants. Properly designed reactor and heat-exchanger configuration can significantly improve productivity of this reaction step.

The second step theoretically needs three moles of sodium hydroxide for every mole of pyrmethyl chloride. However, the patent describes using about 175 percent excess caustic. Although some of this excess is needed to neutralize the excess thionyl chloride used in the first step, it can still be reduced.

The patent suggests that the reaction can be carried out at about 45°C for about two hours. Again, as suggested above, the azeotrope temperature of toluene/water is higher than the suggested reaction temperature, thus it is very feasible to raise the reaction temperature well above 45°C. Raising the temperature would improve the reaction rate considerably. Commercially available heat exchangers and proper process design will improve process productivity. The optimum temperature for the second step can be determined using generally well-known chemical-engineering principles.

Using stoichiometry and solubility of the two steps, it is possible to increase the productivity of the process. Based on the stoichiometry and the process description, I believe that the reactions are zero order. This chemistry is an excellent candidate for a continuous process.

USP 6,875,893 and 7,057,069

As I have indicated, patents reveal significant information as to what the inventors were thinking as they were developing the product and its chemistry. Patent attorneys camouflage the information but still divulge a sufficient amount so that a reader well versed in deciphering can learn and improve on the process or use the information to develop an excellent commercial process.

U.S. patents '893 and '069 are reviewed and interpreted. Following are the reaction steps Figure 5.5 to produce modafinil (a-9benzhydrylsulinyl) acetamide:

- Step 1: benzhydrol + thiourea + HBr → S-benzhydrylthiouronium bromide + H_2O
- Step 2: S-benzhydrylthiouronium bromide + KOH → Potassium Benzhydrylthiol + KBr
- Step 3: potassium benzhydrylthiol + chloroacetamide → 2-(benzhydrylthiol) acetamide + KCl + H_2O (A phase separation is needed at this point)
- Step 4: 2-(benzhydrylthiol) acetamide + acetic acid + H_2O_2 → 2-(benzhydrylsulfinyl) acetamide.

Table 5.4 compares the theoretical stoichiometry against the patent suggested stoichiometry. In a batch process one would expect that more than theoretical amounts of raw materials would be used. The stoichiometry presented is most likely not optimized. Based on my experience, the process conditions and stoichiometry can be optimized for a commercial batch process. Stoichiometry can also be used to calculate the factory cost of the product, as shown in Table 5.5. It can be used to optimize processes and cost.

A review of the operating parameters and chemistry suggests that the reaction and process are simple enough for the product to be easily produced using a continuous process. A schematic of a continuous process is shown in Figure 5.6.

Some of the opportunities for designing a continuous process are discussed. One scheme involves the use of solvents, as suggested in the patents. An alternative is to eliminate the use of any solvent in the first reaction step. In this solution a solvent could be used after the first step.

Solubility of solid raw material reactants in solvents should be considered, as it facilitates process control and reactivity. Process-control

Table 5.4   Modafinil raw material mole ratios.

	Theoretical	Suggested
Benzhydrol	1.0	1.0
Thiourea	1.0	1.2
HBr	1.0	1.2
KOH	1.0	2.9
Chloroacetamide	1.0	1.5
Acetic acid	1.0	2.0
H_2O_2	1.0	2.0

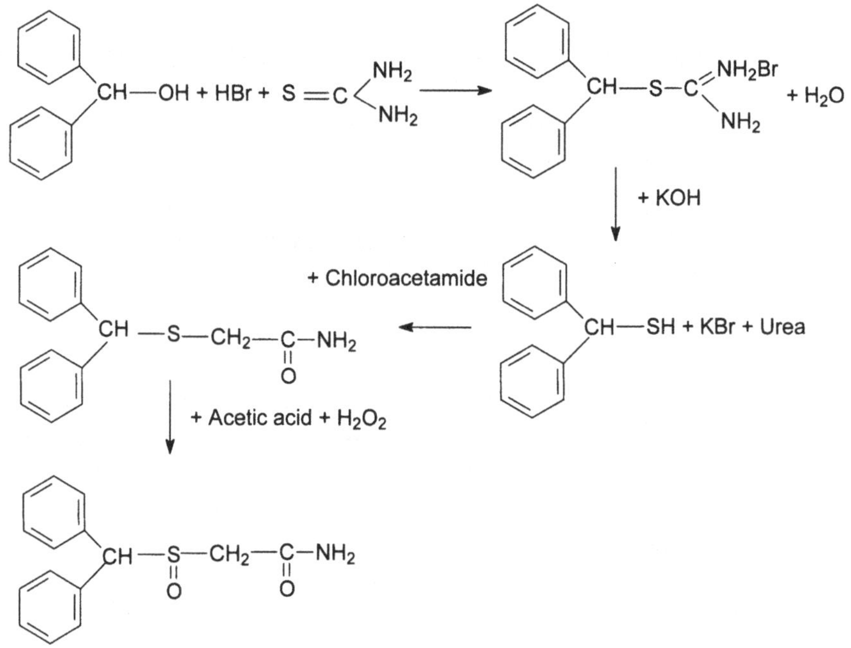

Figure 5.5   Modafinil chemistry.

technologies are more amenable to liquids, making process control economical and efficient. An understanding of the reaction mechanism is important, as it can facilitate the process.

The patents discuss use of THF (tetrahydrofuran)/water, MCB (monochlorobenzene)/water, and other potential solvents. In solvent selection a few parameters need to be reviewed. They are mutual solubility, density, azeotrope temperature, and the need to remove reaction byproducts. etc.

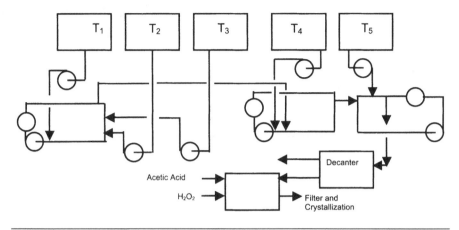

Figure 5.6   Schematic of modafinil continuous process.

Table 5.5   Modafinil cost.

Modafinil cost USP 6,875,893					
	Moles	MW	pounds	price, $/lb	$
Benzhydrol	1	184	83.72	3.18	266.39
Thiourea	1.2	76.1	41.55	1.00	41.55
HBr	1.2	81	44.23	0.80	35.38
KOH	2.88	56	73.38	0.55	40.36
Chloroacetamide	1.5	93.5	63.82	1.30	82.96
Acetic acid	3.2	60	87.36	0.25	21.84
H_2O_2	2	34	30.94	0.27	8.45
MeCl		84	5.00	0.35	1.75
Crystallization Solvent					2.00
					500.68
	Yield	80.40%			
	Product	100		$/lb	5.01
Modafinil		273.35		Conversion expense	3.34
				Total, $/lb	8.34
	Margin	40%	Factory cost	Sell, $/lb	13.91

USP '893 and '069 suggest that the first reaction step should be conducted at about 70°C. However, if THF/water is the solvent system of choice, THF/water azeotrope will start at about 65°C, and the reaction temperature will not get up to the suggested level. MCB/water azeotrope temperature is higher, and the reaction can be run at 70°C. Another alternate solvent system could be toluene/water. This also has high azeotrope temperature. Water alone is also an alternative, but solubility of raw materials and intermediate reaction products need to be considered. USP 4,177,290; 7,186,860; 7,244,865; and 7,345,188 discuss other iterations of the modafinil process.

Mechanisms for the first step have been suggested.[2, 3] It is more likely that benzhydrol reacts with hydrobromic acid to produce bromodiphenyl methane Figure 5.7.

Since bromodiphenyl methane is a liquid, it is quite feasible to use benzhydrol as a molten liquid feed. This, done properly, eliminates or reduces the need for a solvent in the first step. This improves the productivity of this step to a great degree. An understanding of how the chemicals behave and react can allow manipulation of material additions and their feed rates to facilitate the reaction. This is further discussed in Chapter 7.

Mass balance shows stoichiometry of the reactants. Actual mass balance would account for each component going into a chemical reactor and coming out of a reactor or a process. Figure 5.8 is an illustration.

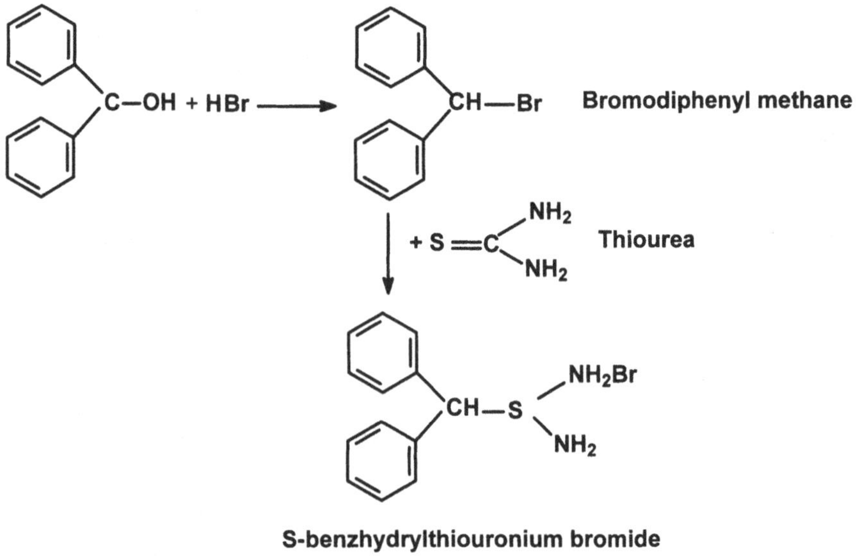

Figure 5.7 Potential mechanism modafinil step 1.

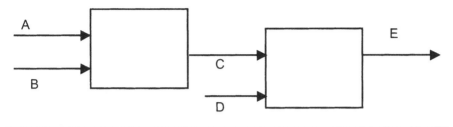

Figure 5.8 Schematic flow diagram.

Mass balance is one of the fundamental components of the chemical-engineering curriculum. Many books have been written on the subject. Each component of streams A to E has to be identified and its quantity accounted for in the mass balance. The basic premise of the mass balance is:

$$input = output$$

Using a reactant as the key component of the reaction system, the yield of the product can be calculated. This information can be used not only for the process-equipment design but also for process simplification and improvements. As the reaction completion rate is not 100 percent, the chemical reaction equation should be balanced with the byproducts produced. Writing and balancing identifies improvement opportunities.

Mass balance should include physical properties and the state of the component. As I have stated, mass balance tells us the state of the process, how it was developed, and the thinking and rationale of the developer and process designer as they scale up a laboratory process to a commercial operation.

HEAT BALANCE

Heat balance suggests the details of the heat that is input or removed from a reactive or a blending process. Heat of a reaction and its formation are very important and an integral part of understanding the chemistry of a reaction. It allows one to ingeniously manipulate the rate of reaction and progression. Heat of reaction of a chemical reaction can be calculated using enthalpies of formation or bond energies. These are very well explained in thermodynamics books.[4] Heat of formation using enthalpies of formation are calculated using the formula:

$$\Delta H \text{ reaction} = \sum H \text{ products} - \sum H \text{ reactants}$$

In the above equation H products and H reactants are enthalpies of reaction products and reactants of a balanced chemical reaction. Heat of formation using bond energies (BE) are calculated using the formula:

$$\Delta H \text{ reaction} = \sum y \text{ BE reactants} - \sum x \text{ BE products}$$

Heat of solution and specific heat of reactants are of value in reactive processes, as they can be used and maneuvered along with the physical properties of the chemicals involved in the reaction to develop and design a cost-effective process.

FINANCIAL VALUE OF MASS AND HEAT BALANCE

Mass balances for reactive and blending processes are the basis of the economics and justification of the processes. They are also used to develop the factory cost and the selling price of the products. They are living documents, and the processes need to be monitored against an established standard. Such monitoring ensures product quality and profitability.

Mass balance needs to done on a daily and monthly basis. Each has a separate function and value. Daily mass balance is of value to operations whereas the monthly mass balance is value to people responsible for financial profit and loss.

If a mass balance for each product is done on a daily basis, it along with product quality will show any deviation from established standards. It will show if the process has any excessive use of any raw material, has lower than standard yield or the quality has deviated from standard. This can be used address any process changes and equipment malfunctions. Table 5.6 is an illustration showing excessive use of hydrobromic acid, potassium hydroxide and hydrogen peroxide. All of the numbers are hypothetical to illustrate value of mass balance.

Comparison of costs against standard cost shown in Table 5.6 shows that due to lower yield and excessive material use, factory cost has gone up by $ 0.79 per pound.

Cumulative mass balance on a monthly basis will show the financial performance of the product. Depending on the company, if the variations from the standard persist, appropriate resources can be allocated to resolve the manufacturing issues.

Mass balance has another value as it can show the financial gains if the raw material usage and the yield could be improved. In the case of modafinil manufacture if thiourea, hydrobromic acid and potassium hydroxide usage could be reduced by 10% (hypothetical number) and the yield improved by 10% (hypothetical number), the factory cost would

Table 5.6 Modafinil cost: Lower than standard yield and excessive use of materials.

Modafinail cost	USP 6875893		Lower than standard yield along with ex- cessive use of raw materials		
	Moles	MW	pounds	price, $/lb	$
Benzhydrol	1	184	89.75	3.18	285.57
Thiourea	1.2	76.1	44.54	1.00	44.54
HBr	1.5	81	59.26	0.80	47.41
KOH	3	56	81.95	0.55	45.07
Chloroacetamide	1.5	93.5	68.41	1.30	88.93
Acetic acid	3.2	60	93.65	0.25	23.41
H_2O_2	2.2	34	33.17	0.27	9.06
MeCl		84	5.00	0.35	1.75
Crystallization Solvent					2.00
					547.75
		Yield	75.00%		
		Product	100	$/lb	5.48
Modafinil		273.35		conversion expense	3.65
			Factory cost	Total, $/lb	9.13
	Margin	40%		Sell, $/lb	15.22
			Loss from standard cost, $/lb		-0.79

be improved by $1.05 per pound. This is significant improvement compared to the cost illustrated in Table 5.5.

Mass balance can be used to discern "what if" values of process optimization and yield improvement. This can be done for reactive as well as blending of chemicals. I used such an exercise for planning as well justification for resources to improve profitability.

REFERENCES

1. Box, G. E., Hunter, W. G., Hunter, J. S., Hunter, W. G., *Statistics for Experimenters: Design, Innovation, and Discovery*, 2nd Edition, Wiley, 2005.

2. Kofod, Helmer. "Furfuryl Mercaptan, an Improved Preparative Method." *ACTA Chemical Scandinavia.* 7;No. 9;1953;1302–1306.

3. Vetter, Stefan. "Direct Synthesis of Di- and Trimethyoxybenzyl Thiols from the corresponding alcohol." *Synthetics Communications.* 28;(17);1998;3219–3223.

4. Dodge, B. F. *Chemical Engineering Thermodynamics.* McGraw Hill Book Company Inc., 1944.

Chapter 6

Reaction Kinetics

Understanding the reaction kinetics of a chemical reaction is extremely important. Once we understand the what, the why, and the how of the kinetics and the impact on the reactants and the intermediates, it is very possible to manipulate as well improve the reaction conversion to achieve higher yield. This understanding, along with the physical properties of the reactants, solvents, and intermediates, facilitates commercialization of the process.

Fundamentals of reaction kinetics have been discussed very thoroughly in various textbooks. However, some of the methods by which reaction rates can be manipulated, improved, and advanced are discussed. Reactions are facilitated if all of the reactants are in liquid phase or dissolved in liquids. Solvents facilitate reactions. In some cases solvents are also the reactant. A catalyst or a promoting agent could be used.

REACTION RATE AND VARIABLES

To reduce cost and improve productivity, our objective is to increase the rate of reaction. It is expected that improvement of the reaction rate will reduce the reaction time; and, since the reaction time is lowered, it will reduce the size of the processing equipment needed and thus the investment. In order to achieve these objectives, it is necessary to recognize the parameters that come into play in every reaction.

Many reactions are reversible. Following are some of the ways to push the reaction forward to improve yield and productivity:

Concentration

Increase the concentration of the reactants in the reaction mass. In batch processes this has to be carefully considered, as side reactions can result in unwanted products, especially when one of the reactants is a solvent. Reaction concentration can be achieved by removing the reaction product either by reacting with another reactant in the subsequent reaction step or by separation/extraction or azeotropic separation.

Physical State

If the reactants are liquids, that simplifies the process. If the reactants are solids, they should be dissolved in a solvent. This facilitates better process control and is especially beneficial for continuous processing. Every effort should be made to solublize the reactants and the product in a single solvent. Two solvents can add to the complexity of the process, but, if the process yield and the productivity are improved, it is worth using two solvents. In such case water and an organic solvent should be preferred. In the selection of solvents, mutual solubility and the method of separation and recovery should be a consideration. Higher mutual solubility can reduce the process yield. Mass transfer of the reactants and products between the two solvents has to be a consideration. Gravity separation or azeotropic distillation should be the preferred choices to separate solvents. Energy conservation should always be a consideration in every process.

Temperature

Raising the reaction temperature usually delivers more energy into the system and increases the reaction rate by causing more collisions between the particles. As a rule of thumb, rates for many reactions double for every 10 degrees Celsius increase in temperature. The effect of temperature may be larger or smaller than the rule of thumb. This should be exploited.

Heat of Reaction

Heat of reaction should be utilized to advantage to improve the reaction rate. By controlling the exotherm, the reaction can often be driven forward and the reaction productivity improved.

Order and/or Method of Addition

Order and method of reactant addition can have a significant influence on any reaction rate, yield, and process stoichiometry. This might not seem to be of value, but if the reactants are added in a different sequence compared to the methods tried in the laboratory, the yield and the raw-material consumption can be greatly improved. If the stoichiometry can be improved, we can have a process that is sustainable and requires reduced investment in waste-handling methods. This also reduces the cost of the product and can improve profitability.

Rate of Additon

Reactant addition rate can have a significant impact on the total reaction rate. Many times it is not feasible to define the rate of addition of reactants in the laboratory. This is especially true in the initial development stages of the product and process. Understanding the thermodynamics and visualizing the unit operation that will be used in the commercial process can define the rate of addition. The laboratory has limitations, and it might not be feasible to test higher feed rates. The impact of rate variation can be tested in pilot plant provided that it has the capability.

Other Parameters

The physical state of the reactants can be helpful. Liquid is the preferred method. Liquids can be metered and solids can be dispensed using weight loss or other similar techniques. Solids can be dissolved, which can improve the process, but dissolution is an extra processing step. If the reactions involve use of a gas, its flow can be best controlled in a liquid form. After the gas has been vaporized, it can be used in the reaction. Laboratory experiments are on a small scale and the raw materials are used in the form they are available from the suppliers. They can be easily handled in the lab. However, on a commercial scale the methodology has to be such that the processing time is minimized.

Applications

Following are some examples of how reaction kinetics can be improved by applying these considerations.

Removal of the reaction product or reacting it with another chemical species can speed up the reaction rate. However, this generally happens in a continuous process. Batch processes generally do not afford these opportunities. For this reason, the yield of multistep batch processes conducted in the same reactor is generally lower than continuous processes. Higher yield can be achieved for a multistep batch process if the byproducts are separated after each step. This option generally results in investing in multiple in-process tanks, thereby increasing cycle time and investment. Economic consideration is necessary for choosing the process type.

DIAZOTIZATION

Let us consider a diazo reaction. In a batch process, a chemist/engineer will diazotize the whole batch at a low temperature to obtain a good conversion. Since diazo compounds are considered unstable, most of the reactions are conducted below 0°C and at high dilution. High dilution is used to dissipate the exotherm. Low temperatures require high-energy consumption. Dilution also lowers batch productivity and increases the waste treatment load.

If the diazo product were reacted with the reactive chemical of the following step as soon as it was produced, the yield would be higher because only stoichiometric reaction will take place. This can happen if the reactions are carried out at room and/or higher temperatures. Carrying out the reaction at a higher temperature (remember that process safety cannot be overlooked) reduces energy consumption. It definitely improves productivity, as the reaction can be carried out at a higher concentration. This can be achieved in a batch as well as a continuous process. However, it is much more convenient in a continuous process. If the diazonium product is not reacted with the intended molecule, it can react to produce diazo tars and other side products, and the yield is lowered.

Commercial batch diazotizations also create another challenge in the plant. Due to an excess of nitrite and/or acid, they generate nitrous oxide, a pollutant, which has to be contained by installing a scrubber.

The order of reactant addition, as stated earlier, is important. In a batch diazo operation the addition order would be amine, acid, and nitrite. If the order is amine, nitrite, and acid, we will have two competing reactions of diazo and nitrous-oxide formation, a product and a pollutant, respectively. The yield of the second addition method will be lower.

Chemists and chemical engineers, who are experienced and familiar with such addition methods, can definitely take advantage of addition methodologies. Processes for methyl anthranilate, metformin hydrochloride, phthalimide, and latex polymers, which are described in other chapters, are a few examples of the exploitation of methods of reactant addition.

REACTANT AS A SOLVENT

Another way to improve the reaction kinetics and the yield is to separate/extract the reaction product from the process. This can prevent reaction of the product with one or more of the reactants to form undesirable byproducts and reduce the yield or the reaction rate. This can be done only in a quasi-batch or continuous process.

Sometimes a reactant is also used as a solvent in a reaction. This is the traditional way to develop a product/process in the laboratory. It is an economical way to conduct a reaction but can reduce productivity and can impede the reaction progress. It can also result in lower yield due to the formation of byproducts. An example of the above is:

$$A + B \rightarrow C.$$

A and B are two liquids that react in the presence of a catalyst to produce product C in the liquid phase. B is used as a solvent for the reaction. Chemical A reacts with B to form the product C, but due to its excess it also produces undesired products, thereby reducing the overall process yield. A is an aqueous organic solution in which the catalyst is dissolved.

Product C is insoluble in water, and it azeotropes with water at a temperature that is close to the boiling point of reactant B. Conventionally reaction is carried out at a temperature that is close to the azeotrope temperature.

The parameters defined above can be used to manipulate the reaction. The reaction temperature can be raised as high as necessary to have a reflux and introduce reactant A with the catalyst in the vapor phase. This will create an artificial excess of B in the vapor space. Product C is azeotroped with water and removed from the reaction mass. Since the product is insoluble in water, we will use two phases. The lower water phase is recycled back to the reactor, making sure that the reactor is not empty.

We were able to improve the yield of the above reaction by about 30 percent over the process whereby an excess of reactant B was used. This option improved not only productivity and yield, but also gave us an environmentally sustainable process. We eliminated the solvent recovery step and were able to minimize the solvent-recovery sludge that was sent to landfill.

During product and process development it is difficult to visualize many manipulations of the physical properties. Incorporating them later in the process creates many opportunities.

OTHER CONSIDERATIONS

Physical properties, the method of addition, and the heat of exotherm have been collectively manipulated to speed up the reaction rate, reduce the cycle time, and improve productivity.

Production of Benzimidazole Compound

Synthesis of a benzimidazole compound is used as a detailed illustration of the use of reaction kinetics. Products of this chemistry are sold as Nexium®, omeprazole, Ocid®, Prilosec®, and other generic and brand names in different parts of the world. The chemistry described in these patents has been known for more than 30 years. Many companies have patents on marginal process improvement nuances. From a pure science viewpoint, it is a synthesis of specialty/fine chemicals that happens to treat a disease.

A review of some of the cited patents[1, 2, 3, 4, 5, 6, 7, 8, 9, 10, 11] suggests that the processes can be operated in a unique and economical manner for the manufacture of 5-methoxy2-(4-methoxy-3, 5-dimethyl-2-pyridinyl) methyl sulfinyl1H-benzimidazole, commonly known as omeprazole and sold as an anti-ulcer agent. Chlorinating an alcohol, which is combined with methoxy-2-benzimidazolethiol to produce an intermediate, is then oxidized to produce the active ingredient. Omeprazole/Nexium chemistry is illustrated in Figure 6.1.

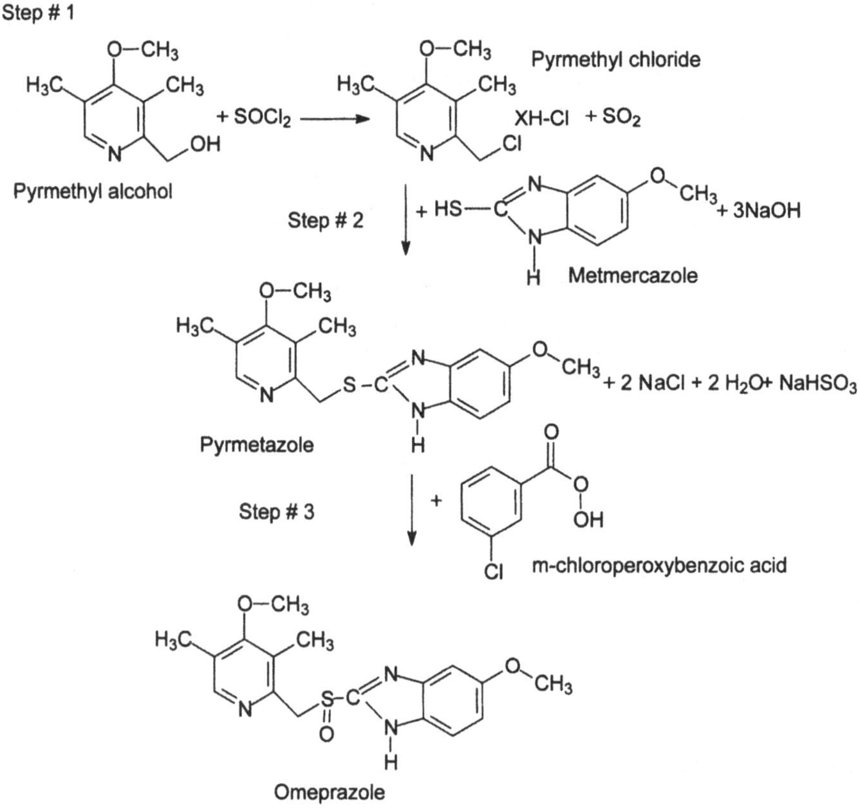

Figure 6.1 Omeprazole/Nexium chemistry.

The combined knowledge of the cited patents and the guidelines outlined above can be used to improve the reaction kinetics. This information can also be used to convert a batch process to a continuous process.

The batch process calls for a solution of pyrmethyl alcohol in a solvent to be reacted with thionyl chloride at ambient temperature. The resulting pyrmethyl chloride was reacted with metmercazole dissolved in an aqueous caustic solution in the presence of a tetrabutylammonium bromide (phase transfer catalyst) at 25 to 40°C. The resulting pyrmetazole is oxidized to omeprazole.

To review the process described in USP 5,958,955 and USP 7,227,024, pyrmethyl alcohol is used as the key component, and every other chemical used is based on it. Toluene is used as the solvent. Use of other solvents is mentioned in the patents. It is necessary to establish solubility of reactants and reaction intermediates to select the appropriate solvent. Solvent recovery and recycling are additional considerations. Table 6.1 compares the stoichiometric ratio of the primary reactants.

If one compares the stoichiometry of the two patents, it becomes obvious that the process outlined in USP '024 suggests the addition of thionyl chloride over a certain amount of time at a certain rate compared to its addition in USP 5,958,955 over a very short time. USP '955 uses less thionyl chloride than '024, thus there is less waste and a greener process, as less alkali would be needed to neutralize leftover thionyl chloride. This suggests that it is feasible to have a continuous process.

USP '024 suggests that use of water-saturated toluene is a better process than that described in USP '955, where thionyl chloride is used as a chloro-hydroxylating agent. However, the thionyl chloride literature suggests that the presence of water should be avoided, as it reacts with water to release hydrogen chloride and sulfur dioxide. In USP '024 and '447[11] wet toluene is used to promote the reaction. This goes against the

Table 6.1 Comparison of Omeprazole/Nexium stoichiometry.

Omeprazole		
USP	7,227,024	5,958,955
	Mole/key	Mole/key
Pyrmethyl alcohol	1.00	1.0000
Thionyl Chloride	1.30	1.1008
Metmercazole	1.03	1.0000
Pyrmetazole	not part of	
Meta Chloro peroxybenzoic acid	the patent	0.9901

conventional wisdom. However, if water improves the conversion, it is worth exploration and commercialization, especially if the results are significantly better than the reaction using dry toluene. The presence of water with thionyl chloride would present the challenge of selecting appropriate construction materials, but it can be easily handled.

The use of wet toluene offers another processing advantage. Since toluene will be recovered and reused, wet toluene can be used. Recovered toluene is saturated with water and has just enough water to meet the requirements in the patent. This simplifies the process. Inventors exploring alternatives to improve yield may be considered controversial, as they are going against the norm. However, if thinking outside the box gives better results, it is innovation.

In USP '024, thionyl chloride is added over an extended time compared to '955, and the conversion rate is about the same. The amount of thionyl chloride being used in '024 is significantly higher (about 18 percent). This excess suggests that faster addition of thionyl chloride improves the reaction rate and should be preferred. Reaction temperatures have been suggested in the patents, but it might be feasible to operate at higher temperatures. Based on my experiences, the process description suggests zero-order reactions. Such reactions are ideal for a continuous process. We all know such processes have high productivity and quality.

Since pyrmethyl alcohol has a low melting point (56 to 61°C), use of molten alcohol should be explored. This can simplify the process. Commercially available technologies can be used to precisely control the process to deliver a quality product. Use of molten alcohol along with proper temperature control can speed the reaction and the conversion.

Since the chemistries are textbook chemistries, it is feasible to use a continuous process whereby the stoichiometry and the solvent selection can be optimized and economized. Solvent selection needs to be carefully done, as the suggested solvents (toluene, methylene chloride) are good solvents for pyrmethyl alcohol, intermediates, and m-chloroperoxy benzoic acid. In addition, they azeotrope with water and are not completely miscible, thereby making separation easier. Phase separation is further accelerated due to their density differences from water. Every effort should be made to maximize the density difference with respect to water, pure solvent, and solution, as higher density differentials will improve the phase separation. Magnesium monoperoxyphthalate[12] and other similar chemicals can be considered as oxidizing agents if the economics justifies.

A continuous process block flow diagram for omeprazole is illustrated in Figure 6.2.

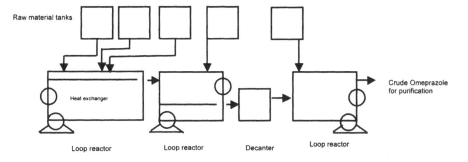

Figure 6.2 Omeprazole continuous process schematic.

USE OF PHYSICAL AND THERMODYNAMIC PROPERTIES

Information generated in the laboratory, along with an understanding of the physical and thermodynamic behavior of the chemicals involved, suggests which unit processes and unit operations can be used to develop a successful and safe commercial process.

Physical and thermodynamic properties of the reactants, intermediates, and solvents can be very effectively used to improve the reaction rate and the process yield. Generally, product development in the laboratory uses traditional methods—i.e., raw materials as solids or dissolved in solvents. Laboratory process and chemistry show us only the feasibility of the reaction. It is not a commercial process but does give us sufficient clues about what the commercial process could look like. Chemists and engineers have to manipulate what is learned in the laboratory to develop a batch or a continuous process that is economical, safe, and environmentally sustainable.

For certain reactions, the reaction mechanism (i.e., how the reactants react) can be used to sequence the addition of the reactants, which in turn improves the reaction rate and yield.

If the reactants and the product can be safely handled in molten form, this can be of advantage, as it reduces the use of a solvent, thereby improving productivity, reaction rate, and yield. Lack of or minimum use of organic solvents significantly reduces energy consumption so that the process can be more environmentally sustainable.

If a solvent is to be used in a process, then its azeotropic relationship, solubility, and density difference with water can be exploited to improve the reaction rate and yield. Solubility of reactants and reaction products can also be exploited.

Heat is the manifestation of the release of energy from a chemical reaction. If this energy can be controlled and manipulated, it can be used to improve the reaction rate. Heat can be removed by using any of the following methods:

- Use of a heat exchanger
- Evaporation of a solvent used in the reaction
- Addition of a cold fluid to absorb the heat of the reaction.

Some examples of reactions that take advantage of either of the above scenarios to improve the yield and reduce batch cycle time have been discussed in other chapters. Chemists and chemical engineers come across many reactions that are exothermic, and they should use component physical properties to improve the process.

EQUIPMENT

A variety of equipment and process control technologies are used in the manufacture of chemicals. Different chemical syntheses require that some of these methods be creatively modified, manipulated, and improved to result in an economical, safe, and sustainable process, whether batch or continuous.

Basically, a reactor is a space where the components react and any heat of reaction is transferred to complete the reaction safely. Traditionally, batch reactions are carried out in jacketed or pressure reactors. These reactors could also be used for a continuous process. Efficient heat transfer propagates the reaction and, based on the residence time and heat transfer, a high conversion yield can be achieved. Plate and frame heat exchangers have been in existence for over 40 years and, due to their high heat-exchange efficiency are ideal for continuous and/or batch processes.

In the last few years microreactors have been touted as technologies to convert specialty and fine chemical (including active ingredient) manufacturing from batch to continuous processing. That might be true for some special reactions, and microreactors should be considered in those cases. However, the economics of various reactor configurations and options needs to be carefully reviewed. Microreactors are generally much more expensive than other reaction options, excluding conventional reactors. The cost-benefit ratio must be the main consideration for their selection.

REFERENCES

1. U.S. Patent 4,045,563. August 30, 1977.
2. U.S. Patent 4,255,431. March 10, 1981.
3. U.S. patent 5,386,032. January 31, 1995.
4. U.S. patent 5,391,752. February 21, 1995.
5. U.S. Patent 5,958,955. September 28, 1999.
6. U.S. Patent 6,166,213. December 26, 2000.
7. U.S. Patent 6,229,091. May 8, 2001.
8. U.S. Patent 6,303,788. October 16, 2001.
9. U.S. Patent 6,919,459. July 19, 2005.
10. U.S. Patent 7,227,024. June 5, 2007.
11. U.S. Patent 2008/0004447. January 3, 2008.
12. U.S. Patent 5,391,752. February 21, 1995.

Chapter 7

Physical Properties

Physical properties play a critical part in the processing of chemicals. They might not seem to have any influence, but they are the heart and soul of any chemical reaction, purification step, and blending operation. They make things happen. Once we understand and master the physical properties, we can manipulate them and simplify processes.

Physical properties of individual raw materials, intermediates, and final products (pure chemical or blend) are unique due to their chemical structures. These characteristics define their specifications and should be exploited to develop a process that will deliver quality product for the desired performance. Deviation outside the desired specification range is a manifestation of process shortcomings. These specifications can be used to simplify or improve existing and develop new processes.

Organic chemicals have many physical properties, but the following are critical. They play a major role in simplifying and facilitating a process. Other properties should not be excluded, as they might have value in certain processes. These critical properties are:

- Physical state
- Density
- Melting/freezing point
- Boiling point
- Azeotrope characteristics
- Solubility
- Heat of reaction
- Heat capacity
- Viscosity.

These properties, when exploited individually and/or collectively, facilitate the processing and blending of chemicals.

PHYSICAL STATE

The physical state of chemicals is extremely important, as it determines how the flow of the chemical will be managed. Its control manages process stoichiometry. Chemicals can be in solid, liquid, or gas form.

For a chemical process, the order of preference for the physical state should be liquid, solid, and gas. For flow control and measurement, liquids are the easiest to measure and handle, followed by solids and gases. Solid feed can be handled through various methods, such as weight-loss feeders or rotary valves. When handling gas, it is preferable to measure the flow as a liquid and then vaporize it for use in the reaction. Gas at times can enhance the mixing process when it bubbles in the liquid and improves the mixing.

Solvents are an important part of a chemical reaction. They facilitate various unit operations and unit processes. Solubility of components dictates the amount of solvent use. Since solvents can take up a large volume of the process vessel, it is necessary to minimize their volume and the number used in a process. Minimization reduces capital-equipment investment in storage tanks, recovery systems, and disposal of distilled bottom solids. Minimization of the number of solvents used also lowers the cost of manufacturing through volume purchase discounts.

Solids that are soluble in solvents are easy to handle and feed in the reaction and/or blending processes. However, solids that do not dissolve can only be added as slurry or by dumping them through a properly designed system to protect personnel and the environment. Feed of corrosive and hazardous solid additions can be a challenge. If such solids can be slurried and their flow controlled, a safe and controlled process is possible. However, the solids, when suspended in a liquid, have an inherent property of settling as soon as the agitation is stopped. This can make uniform feed of a solid suspended in a liquid a challenge.

Various dispersing agents have been invented to retain such suspensions. They work well and facilitate the feeding process. Solids that are corrosive and hazardous in nature always demand extra precaution. Reactive solids can be a challenge.

An excellent example would be the use of aluminum-chloride slurry in Friedel-Crafts reaction. Uniformly suspended slurry can facilitate many reactions. If aluminum chloride can be dispersed using a dispersing agent that is inactive with the solvent and the reaction mass, it can facilitate the process. In most instances it would be preferable to remove the dispersing agent from the reaction mass with the waste stream after the reaction so that it does not contaminate the product. This would require that the dispersing agent be soluble in water. Such finessing of the physical properties can simplify a reaction and make it safe and environmentally sustainable.

DENSITY

Density of liquids is important, as it can be exploited to separate two immiscible liquids. The higher the density difference, the better and quicker the separation of the two liquids for improved processing. Density also changes due to solubility of organic and inorganic chemicals that dissolve in water and organic solvent. This can be used as an advantage. Solubility and density differences can be capitalized to separate two liquids in a gravity decanter. Decanter design is discussed in Chapter 4.

MELTING/FREEZING POINT

Melting/freezing point might not look like an important physical property for an organic chemical. However, if a chemical is a solid, the melting and freezing points matter. If the raw material can be handled in molten form and safely transported, it can be stored at a customer's site in temperature-controlled heated tanks. The molten liquid can be pumped and metered to a process to facilitate it.

With respect to intermediates produced, melting point also becomes important. If the process temperature can be maintained above the melting point, it facilitates processing. Liquid intermediates can reduce the amount of solvent needed for the process. Homogeneous liquid mass facilitates handling, mixing, and any other processing.

Molten liquids can also be used as a heat-transfer medium to control the processing conditions and improve the reaction rate.

BOILING POINT

Boiling point, like melting point, is an important property. It can be used in multiple ways by the process developer to manipulate the reaction. If the reaction is exothermic and sufficient heat is generated to raise the temperature of the mixture above the boiling point of one of the reactants, byproducts, and/or solvents, we can allow the solvent/chemical to boil and condense back to the reactor, thereby partially removing the heat of the reaction from the system. This speeds the reaction and moves it to completion. Chemical evaporation and condensation can reduce the heat transfer load and thus equipment investment. The boiling point of the reaction byproducts can be used for their removal and separation and should be considered in the reaction scheme.

Boiling point at times can also act as a reaction temperature limiter. If the reaction temperature needs to be higher than the boiling point to speed up the reaction, it might be necessary to select a solvent that has the appropriate temperature to push the reaction forward. Economics, solubility, and density differences come into play and should be part of the total equation.

AZEOTROPE CHARACTERISTICS

Azeotrope is an excellent physical property that should be exploited as much as possible. In the reaction step, it can be used to remove the heat of reaction by azeotroping the solvent/chemical, recycling it, and removing a reaction byproduct if it is formed. It can also be used to remove water and/or product or recycle one of the reacting chemicals back to the reactor. It can also be used for solvent recovery and purification. An example of how to exploit this property is discussed in Chapter 2.

It is a useful property for product purification. Solvents that have a lower boiling azeotrope temperature than either of the components facilitate processing.

SOLUBILITY

Solubility of reactants and reaction products in solvents and mutual solubility of solvents are of value in every chemical reaction. Solubility of reactants in solvents that are solids can be used to control the feed, thereby the reaction rate in a batch and/or continuous process.

Solubility of the reaction intermediates and products is of value in enhancing the reaction rate. Raising the temperature of the liquid mass can raise solubility. For reactive systems, raising the temperature can also raise the reaction rate, which benefits the whole process.

Mass transfer plays a role in promoting the reaction. Solubility is also of value in purification, crystallization, spray drying, and other unit operations. Solubility is the basis for crystallization. As the solutions saturate with the evaporation of the solvent, crystals grow, leading to separation of the crystallized product. Besides cost, solubility of reaction byproducts in a chemical reaction should be considered in raw-material selection, as it can simplify product purification and processing.

In reactions using water and an organic compound as a solvent, it is important to select solvents that have minimum mutual solubility. This property becomes important for solvent recovery and reuse.

Along with the above benefits, solubility has a negative impact on the reaction process if the reaction intermediates are separated and purified. They result in yield loss ranging from 5 to 20 percent, depending upon the solubility.

Solubility is also of value for products that rely on a blended mix of certain chemicals to achieve certain application and performance characteristics. It can reduce processing time.

HEAT OF REACTION

Heat of reaction is a tool that can be and should be exploited. I look at it as a gift to move the reaction forward. It is also a tool to speed up the reaction rate. As the temperature of the reactant rises, it pushes the reaction forward. Safety is an important consideration when using heat to improve the reaction rate.

HEAT CAPACITY

Heat capacity is an important feature of the reaction mass. Reaction components that have higher specific heat capacity require higher heat to increase their temperature compared to components with lower heat capacity. This might not look important, but it relates to the total energy expended in any reaction. For energy conservation specific heat of reaction components should be a consideration.

VISCOSITY

Viscosity of fluids influences processing temperatures and equipment. We all know that increasing temperature can lower viscosity, but it might not be beneficial for temperature-sensitive materials. Thus, careful attention is needed to how to use this property in the process. Solvents can be used to lower the viscosity. While lowering viscosity by dilution, finesse is needed. Viscose material should be added to the solvent, as it will speed dispersion rather than vice versa.

COMBINED VALUE OF PHYSICAL PROPERTIES

Not all of the above properties might seem to be individually important, but collectively they have a significant impact on the total process and its unit operations and thus on process economics.

Chemicals have a unique structure, a distinct fingerprint that we call specification. Any process that is not able to deliver a product with the established specification range is an indication that the process has shortcomings and needs to be improved and simplified. In the development of a product it is important that, in addition to reviewing how the solvents will be recovered and re-used, all of the above-mentioned properties be collectively considered in the development of an overall process.

In Chapter 4, the laboratory process for the production of modafinil was discussed in terms of U.S. Patents 6,875,893 and 7,057,069. They illustrate one of the methods used. There are many other patents describing the manufacture of modafinil (USP 7,244,865; 6,649,786; and 7,057,068). They can be adapted in many different ways. The following example is an illustration and suggestion of how the physical properties can be used to simplify the process.

Figure 7.1 is an illustration of the described chemistry. Following is an edited version of the laboratory synthesis from USP 6,875,893. It can be modified, scaled up, and commercialized.

To a suspension of benzhydrol and thiourea in tetrahydrofuran/water an aqueous 48 percent HBr solution was added over a 10-minute period. During the addition, the reaction mixture was heated to 70°C. After three hours stirring at 70°C, the uronium intermediate was hydrolyzed by addition of an aqueous 9.3N potassium-hydroxide solution over a 55-minute period. After 1.5 hours stirring at 70°C, chloroacetamide in a tetrahydrofuran/water solution was added over 15 minutes. After one

---

Step # 1

Benzhydrol + Thiourea + HBr → S-Benzhydrylthiouronium bromide + H_2O

Step # 2

S-Benzhydrylthiouronium bromide + KOH → Benzhydrolthiourea + KBr + Urea

Step # 3

Benzhydrolthiourea + Chloroacetamide + Alkali → 2-(Benhydrylthiolacetamide) + KCl

Step # 4

2-(Benhydrylthiolacetamide) + Acetic Acid + H_2O_2 → 2- (Benhydrylsufinyl) acetamide [Modafinil]

Figure 7.1 Modafinil chemistry.

hour stirring at 70°C, the reaction mixture was cooled down to 55°C and the stirring was stopped. The lower aqueous phase was removed, and the reaction mixture was again stirred. Acetic acid was added. A 30 percent solution of hydrogen peroxide was slowly added over 30 minutes. After one hour stirring, the reaction mixture was cooled to 20°C, and water was added. The resultant suspension was stirred at 0°C overnight. The suspension was then filtered, and the solid was washed with water and dried to yield modafinil (47.9 grams, 80.4 percent solution). The crude modafinil was purified by recrystallization in methanol.

The first step of this chemistry suggests two potential mechanisms. The mechanism[1, 2, 3, 4] outlined in Figure 7.2 seems preferable and was used to extrapolate how physical properties can be used to simplify the process.

Bromodiphenyl methane, the intermediate formed by the reaction of benzhydrol and hydrobromic acid, is a liquid. Formation of a liquid intermediate facilitates the reaction, and it also gives additional liquid to the reaction mass. Higher reaction temperatures promote the formation of the product, and this reaction can be conducted at a higher temperature (USP '865) than suggested in USP '893.

Physical properties of the reaction components and intermediates are outlined in Tables 7.1 and 7.2.

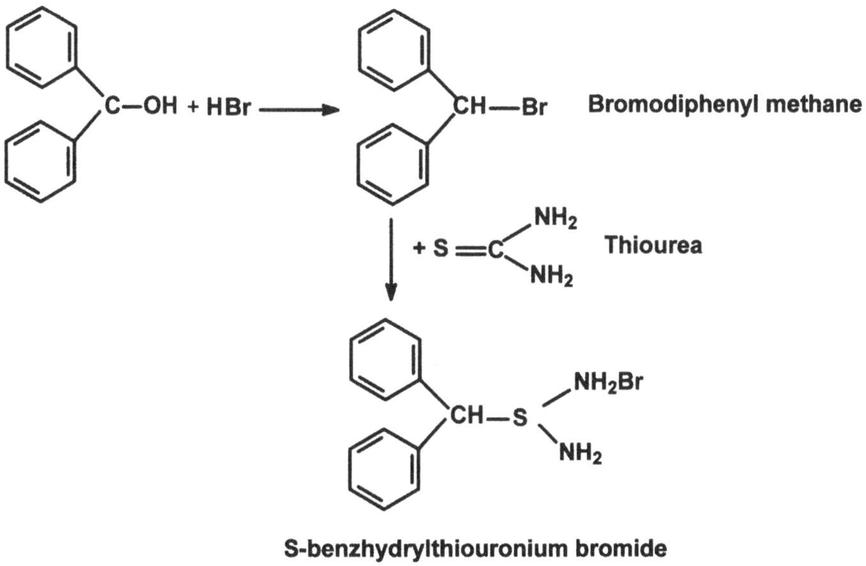

Figure 7.2   Potential route for S-benzhydrylthiouronium bromide.

Table 7.1 Physical properties of Modafinil raw materials.

Chemical	Physical State	MP °C	Solubility
Benzhydrol	Solid	69	
Thiourea	Solid	181	Water
HBr	Liquid		
KOH/K2CO3	Liquid/Solid		
Chloroacetamide	Solid	118	Water
Acetic Acid	Liquid		
H_2O_2	Liquid		

Table 7.2 Physical properties of Modafinil intermediates.

Chemical	Physical State	MP °C
Bromodiphenyl Methane	Liquid	
Benzhydrylthiol	Solid	
S-Benzhydrylthiouronium bromide	Solid	
2-(Benzhydrylthiol) acetamide	Solid	110
2- (Benhydrylsulfinyl) Acetamide	Solid	184

Based on the melting point of benzhydrol, it can be used as a melting liquid. Other raw materials are water-soluble. This suggests that we can eliminate use of a solvent in the reaction system. Lack of solvent also improves the productivity of the process. Solubilities of the intermediates in water are not known, but they are soluble in solvents. If a solvent were needed, my preference would be to use one that will dissolve the intermediates and the final product and be immiscible in water. This will facilitate phase separation, and the solvent could be used for purification and crystallization.

We all have read in our textbooks that the reaction mass should be dry, especially when using chlorinating agents. The recent U.S. Patent 7,227,024 suggests the use of water-wet toluene in chlorination to achieve high yield. Basically, this suggests that what we have been taught to be age-old practice can be challenged and have surprisingly good results. Use of wet toluene facilitates processing, as the azeotroped toluene from the toluene/water mix can be recycled without any additional drying, thereby lowering processing cost.

Raising the reaction mass concentration should always be a consideration. Raising the concentration might require raising the mass temperature, which in turn improves the reaction rate. This is very beneficial for a reactive process, as it improves productivity, reduces cost, and in turn improves profitability. Raising the concentration also reduces waste and associated investment.

EVALUATE ALTERNATE ROUTES

Many chemical syntheses can be done using different routes. Production of o-fluorotoluene sulfonamide,[5] a classical chemistry, can be produced by using any of the three routes shown in Figure 7.3, for example.

These routes can be practiced using either a batch or a continuous process. Since the starting o-fluorotoluene is a liquid, one can capitalize on this property and reduce the amount of solvents used. O-fluorotoluene sulfonyl chloride is also a liquid. With raw materials and intermediates being liquid and the solids being soluble in solvents, a very green process can result. Along with water, use of a single solvent is a distinct possibility. The unit processes are such that they can be easily practiced with standard unit operations, resulting in an excellent continuous process.

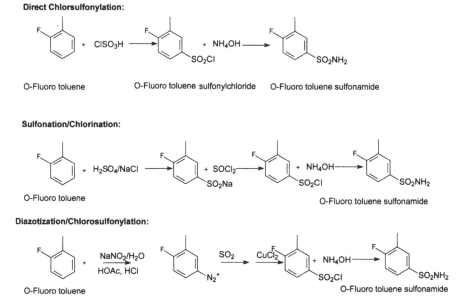

Figure 7.3 Alternate routes for O-Fluorotoluene sulfonamide.

Batch processing is also viable. In either process, isolation of intermediates is not necessary. Process productivity can be improved by raising the concentration. Physical properties can assist. Product demand and economics will dictate the best process.

Table 7.3 suggests the optimization opportunities that exist for the chemistry in USP '487. According to the patent, if the intermediates are not isolated, the yield of the sulfonamide via the first route is about 87.7 percent. The patent indicates the yield for the sulfonyl chloride and the sulfonamide each at about 95.2 percent. When the two products are isolated, the combined yield drops to about 72 percent. These are laboratory yields. If the commercial process is optimized, the yield could be improved to the high 90s in percentile range. We would then have a quality-by-design and sustainable process.

The purpose of the above illustrations is to show that one can use the physical properties of raw materials and intermediates to simplify unit operations, resulting in a process that produces quality product and is sustainable. Since liquids facilitate different unit processes and unit operations, the value of this physical state of a component should be capitalized whenever possible.

Collecting physical properties of various chemicals used in the reaction and blending processes can be a challenge. Raw-material suppliers should facilitate. There are many other sources and books. Resources 6, 7, 8 on the Internet are helpful and should be explored.

Discussion in this chapter illustrates that the physical properties of each reactant and chemical have to be explored and exploited to every reactive and blending process. Examples of such exploitation have been discussed throughout this book. It is the responsibility of every chemist and chemical engineer to venture out and experience, as we are not taught such exploitive methods as part of our education.

Table 7.3   Comparative yields for USP '487.

	Yield, % (Isolations)	Yield, % (No Isolations)
Step #1 (w/o isolation)	95.2	
Step # 1 (isolation)	75.7	
Step # 2	95.2	
Combined (w/o isolation)	90.6	
Combined (isolation)	72.1	87.7%

REFERENCES

1. Kofod, Helmer. "Furfuryl Mercaptan." *ACTA Chemica Scandinavica.* 7;1953;1302–2306.
2. Johnson T. B., and Sprague, J. M. "A New Method for the Preparation of Alkyl Sulfonyl Chloride." *Journal of American Chemical Society.* 58;1936;1348–1352.
3. Johnson T. B., and Sprague, J. M. "The Preparation of Alkyl Sulfonyl Chlorides from Isothioureas." *Journal of American Chemical Society.* 59;1937;1837–1840.
4. Frank, R. L., and Smith, P. V. J. *Journal of American Chemical Society.* 68;1946;2103–2104.
5. USP 7,579,487. August 25, 2009.
6. Chemblink. Accessed September 18, 2009.
7. Chemspider. Accessed September 18, 2009.
8. NIST. Physical & Chemical Properties of Chemicals. Accessed September 22, 2009.

Chapter 8

Combination of Heat, Mass Balance, and Physical Properties

The technical operating manual is one of the two living documents of every process. Preparation of the technical operating and process design manuals starts when the development process begins in the laboratory. These are living documents, and they are continually refined as the product and process methods are developed. The operating manual is an excellent tool for learning the process and training operating and maintenance personnel, chemists, and chemical engineers. The process design manual documents the logic and design details of why and how of each piece of equipment that is on the process and instrumentation diagram.

TECHNICAL OPERATING MANUAL

As explained earlier the technical operating manual includes how to operate the process and the basis of the process. It should include each of the following items as a chapter. To protect confidentiality, in some cases in the example entries below the chemicals are not identified by name. In addition, instead of describing an actual process, I have used examples for different chemical syntheses that are in the public domain. Each company can use its own discretion to add, modify, and/or delete any chapters to suit their needs. Here are the elements:

1. **Process description:** This briefly describes each reaction step and what each step entails. It also describes the process conditions of each step.

2. **Unit charge and material recovery sheet:** This includes the weight and molar basis of each reactant used in the process. It should also include each reaction product.

3. **Raw-material specification and material safety data sheets (MSDS):** Each of the raw materials used in the process and its specifications. This section should also include the MSD sheets for each raw material.

4. **Process chemistry:** This section details the process chemistry. The reaction steps and by-products should be included. The chemical equation should be balanced, as it will show the developer's understanding of the process chemistry.

5. **Heat and mass balance:** This details the mass balance for the commercial process. This is used for process-equipment design. Heat balance is also used in equipment design.

6. **Process conditions and effects of variables:** This section details the desired operating conditions. This section also includes information as to how changes in operating conditions can affect process yield and the effect of variables on the process.

7. **Equipment description and process and instrumentation diagram:** This defines the use of the equipment used in the process. A diagram is very beneficial, as it details the actual plant equipment and piping.

8. **Suggested operating procedure:** This includes how the actual process should be operated. Based on my experience, this document should be developed by the process developers but should be reviewed and modified if necessary by the operating personnel to ensure that the process is operated in the best manner. This can be used for training operating personnel.

9. **Laboratory synthesis procedure:** Laboratory synthesis details how the process was carried out in the laboratory by the chemist who developed the final process that has been commercialized. This is a very helpful section, as it can be used in the future to further simplify the synthesis and blending operation.

10. **Additional critical notes:** This includes any other information that is important for the process. This section can include process startup and shutdown procedures. Safety should be a prime consideration during startup and shutdown.

11. **Analytical methods:** This details each in-process test that should be used to check in-process and final product quality.

12. **Thermodynamic and physical property data:** This section includes properties of each raw material and intermediate. Having this data is very important, as it can be used for better understanding of the process and for expansion if so desired.

13. **Troubleshooting guide:** This is helpful to solve operating problems if and when encountered.

PROCESS DESCRIPTION

Resin XYZ is produced by reaction of appropriate substituted phenol with formaldehyde in the presence of an acid catalyst. The mole ratio of the formaldehyde to the substituted phenol is 0.87. All of the substituted phenol is charged to the reactor. Formaldehyde is added, and the acid catalyst is charged. The batch is heated to 80°C. The exotherm of the batch carries the reaction forward. Once the reflux stops, the reminder of the catalyst is added. Ring and ball temperature of the resin is monitored. Once it is achieved, the acid catalyst is neutralized with an amine. Molten product is flaked and bagged.

UNIT CHARGE AND MATERIAL RECOVERY SHEET

Table 8.1 shows the batch charge. It is for 1000 pounds of the product.

RAW MATERIALS SPECIFICATION AND MATERIAL SAFETY DATA SHEETS (MSDS)

Raw material specs (Figure 8.1) for benzhydrol is a typical example. Such information for each raw material and solvent should be included. In the last few years MSDS are generally used for the technical data sheets, as MSDS do not have sufficient information. Data includes solubility, azeotrope, or specific heat data.

Material safety data sheet for each raw material, intermediate (if isolated), and final product should be included. MSD sheets for benzhydrol (Figure 8.2) and 1,3-diphenyl-2-thiourea (Figure 8.3) are included as examples.

PROCESS CHEMISTRY

One needs to describe the chemistry of the each reaction step, its operating conditions. This should also discuss the byproducts that are produced. This is a quick description of the process. Detailed chemistry description is very helpful. For reactive processes, it is important that each reaction step be written as a complete balanced equation. Following is an example.

$$NaOH + HCl \rightarrow NaCl + H_2O$$

Alfa Aesar

Shore Road, Port of Heysham Industrial Park
Heysham, Lancashire LA3 2XY
United Kingdom

Tel +44 (0) 1524 850506   www.alfa.com
Fax +44 (0) 1524 850608

Product Specification

Catalogue Number:	**A12884**	**Alternative No :**	**L03079**
Product Name:	**Benzhydrol, 99%**		

Alternative Name:	Diphenylcarbinol / Diphenylmethanol

Structure:

Chemical Abstract No:	91-01-0
EINECS:	202-033-8
TSCA:	Listed
Molecular Formula:	C13H12O
Molecular Weight:	184.24

TYPICAL DATA
(Technical information only)

Description	White crystalline powder
Melting Point:	64-68℃
Boiling Point:	297-298℃
Density:	No data
Flash Point:	160℃

	Specification	**QC Method**
Assay(GC)	>98.5%	AS2

Hazards :	
Storage :	Store in well sealed containers, under cool, dry conditions.

Approved By:	Dr. Eric Cuthbertson	**Retest :**	7 Years
	Technical Services Manager		
Date of Issue:	03 March 2009	**Issue Status**	2

Figure 8.1   A raw material specification for benzhydrol.

Health	2
Fire	1
Reactivity	0
Personal Protection	E

Material Safety Data Sheet
Benzhydrol MSDS

Section 1: Chemical Product and Company Identification

Product Name: Benzhydrol

Catalog Codes: SLB1867

CAS#: 91-01-0

RTECS: DC7452000

TSCA: TSCA 8(b) inventory: Benzhydrol

CI#: Not available.

Synonym: alpha-Phenylbenzenemethanol

Chemical Name: Benzhydrol

Chemical Formula: C13-H12-O

Contact Information:

Sciencelab.com, Inc.
14025 Smith Rd.
Houston, Texas 77396

US Sales: **1-800-901-7247**
International Sales: **1-281-441-4400**

Order Online: ScienceLab.com

CHEMTREC (24HR Emergency Telephone), call:
1-800-424-9300

International CHEMTREC, call: 1-703-527-3887

For non-emergency assistance, call: 1-281-441-4400

Section 2: Composition and Information on Ingredients

Composition:

Name	CAS #	% by Weight
Benzhydrol	91-01-0	100

Toxicological Data on Ingredients: Benzhydrol: ORAL (LD50): Acute: 5000 mg/kg [Rat]. DERMAL (LD50): Acute: >5000 mg/kg [Rabbit].

Section 3: Hazards Identification

Potential Acute Health Effects: Hazardous in case of skin contact (irritant), of eye contact (irritant), of ingestion, of inhalation.

Potential Chronic Health Effects:
CARCINOGENIC EFFECTS: Not available.
MUTAGENIC EFFECTS: Not available.
TERATOGENIC EFFECTS: Not available.
DEVELOPMENTAL TOXICITY: Not available.
Repeated or prolonged exposure is not known to aggravate medical condition.

Section 4: First Aid Measures

Eye Contact:
Check for and remove any contact lenses. In case of contact, immediately flush eyes with plenty of water for at

---

Figure 8.2   An MSDS file for benzhydrol.

Odor Threshold: Not available.

Water/Oil Dist. Coeff.: Not available.

Ionicity (in Water): Not available.

Dispersion Properties: Not available.

Solubility: Very slightly soluble in cold water.

Section 10: Stability and Reactivity Data

Stability: The product is stable.

Instability Temperature: Not available.

Conditions of Instability: Excess heat, dust generation, incompatible materials

Incompatibility with various substances: Reactive with oxidizing agents, acids.

Corrosivity: Non-corrosive in presence of glass.

Special Remarks on Reactivity: Incompatible with oxidizing, acid chlorides, acid anhydrides, and acids.

Special Remarks on Corrosivity: Not available.

Polymerization: Will not occur.

Section 11: Toxicological Information

Routes of Entry: Inhalation. Ingestion.

Toxicity to Animals:
Acute oral toxicity (LD50): 5000 mg/kg [Rat].
Acute dermal toxicity (LD50): >5000 mg/kg [Rabbit].

Chronic Effects on Humans: Not available.

Other Toxic Effects on Humans: Hazardous in case of skin contact (irritant), of ingestion, of inhalation.

Special Remarks on Toxicity to Animals: Not available.

Special Remarks on Chronic Effects on Humans: Not available.

Special Remarks on other Toxic Effects on Humans:
Acute Potential Health Effects:
Skin: Causes skin irritation.
Eyes: Causes eye irritation.
Inhalation: Causes respiratory tract and mucous membrane irritation.
Ingestion: May cause gastrointestinal tract irritation.
The toxicological properties of this substance have not been fully investigated.

Section 12: Ecological Information

Ecotoxicity: Not available.

BOD5 and COD: Not available.

Figure 8.2 An MSDS file for benzhydrol. *(continued)*

Use a shovel to put the material into a convenient waste disposal container. Finish cleaning by spreading water on the contaminated surface and allow to evacuate through the sanitary system.

Section 7: Handling and Storage

Precautions:
Keep away from heat. Keep away from sources of ignition. Empty containers pose a fire risk, evaporate the residue under a fume hood. Ground all equipment containing material. Do not ingest. Do not breathe dust. Wear suitable protective clothing. In case of insufficient ventilation, wear suitable respiratory equipment. If ingested, seek medical advice immediately and show the container or the label. Avoid contact with skin and eyes. Keep away from incompatibles such as oxidizing agents, acids.

Storage: Keep container tightly closed. Keep container in a cool, well-ventilated area. Do not store above 25°C (77°F).

Section 8: Exposure Controls/Personal Protection

Engineering Controls:
Use process enclosures, local exhaust ventilation, or other engineering controls to keep airborne levels below recommended exposure limits. If user operations generate dust, fume or mist, use ventilation to keep exposure to airborne contaminants below the exposure limit.

Personal Protection:
Splash goggles. Lab coat. Dust respirator. Be sure to use an approved/certified respirator or equivalent. Gloves.

Personal Protection in Case of a Large Spill:
Splash goggles. Full suit. Dust respirator. Boots. Gloves. A self contained breathing apparatus should be used to avoid inhalation of the product. Suggested protective clothing might not be sufficient; consult a specialist BEFORE handling this product.

Exposure Limits: Not available.

Section 9: Physical and Chemical Properties

Physical state and appearance: Solid. (Crystalline solid.)

Odor: Not available.

Taste: Not available.

Molecular Weight: 184.24 g/mole

Color: Not available.

pH (1% soln/water): Not available.

Boiling Point: 298°C (568.4°F)

Melting Point: 66°C (150.8°F)

Critical Temperature: Not available.

Specific Gravity: Not available.

Vapor Pressure: Not applicable.

Vapor Density: Not available.

Volatility: Not available.

Figure 8.2 An MSDS file for benzhydrol. *(continued)*

Odor Threshold: Not available.

Water/Oil Dist. Coeff.: Not available.

Ionicity (in Water): Not available.

Dispersion Properties: Not available.

Solubility: Very slightly soluble in cold water.

Section 10: Stability and Reactivity Data

Stability: The product is stable.

Instability Temperature: Not available.

Conditions of Instability: Excess heat, dust generation, incompatible materials

Incompatibility with various substances: Reactive with oxidizing agents, acids.

Corrosivity: Non-corrosive in presence of glass.

Special Remarks on Reactivity: Incompatible with oxidizing, acid chlorides, acid anhydrides, and acids.

Special Remarks on Corrosivity: Not available.

Polymerization: Will not occur.

Section 11: Toxicological Information

Routes of Entry: Inhalation. Ingestion.

Toxicity to Animals:
Acute oral toxicity (LD50): 5000 mg/kg [Rat].
Acute dermal toxicity (LD50): >5000 mg/kg [Rabbit].

Chronic Effects on Humans: Not available.

Other Toxic Effects on Humans: Hazardous in case of skin contact (irritant), of ingestion, of inhalation.

Special Remarks on Toxicity to Animals: Not available.

Special Remarks on Chronic Effects on Humans: Not available.

Special Remarks on other Toxic Effects on Humans:
Acute Potential Health Effects:
Skin: Causes skin irritation.
Eyes: Causes eye irritation.
Inhalation: Causes respiratory tract and mucous membrane irritation.
Ingestion: May cause gastrointestinal tract irritation.
The toxicological properties of this substance have not been fully investigated.

Section 12: Ecological Information

Ecotoxicity: Not available.

BOD5 and COD: Not available.

Figure 8.2   An MSDS file for benzhydrol. *(continued)*

Products of Biodegradation:
Possibly hazardous short term degradation products are not likely. However, long term degradation products may arise.

Toxicity of the Products of Biodegradation: The product itself and its products of degradation are not toxic.

Special Remarks on the Products of Biodegradation: Not available.

Section 13: Disposal Considerations

Waste Disposal:
Waste must be disposed of in accordance with federal, state and local environmental control regulations.

Section 14: Transport Information

DOT Classification: Not a DOT controlled material (United States).

Identification: Not applicable.

Special Provisions for Transport: Not applicable.

Section 15: Other Regulatory Information

Federal and State Regulations: TSCA 8(b) inventory: Benzhydrol

Other Regulations: Not available.

Other Classifications:

WHMIS (Canada): Not controlled under WHMIS (Canada).

DSCL (EEC):
R36/38- Irritating to eyes and skin.
S2- Keep out of the reach of children.
S46- If swallowed, seek medical advice immediately and show this container or label.

HMIS (U.S.A.):

  Health Hazard: 2

  Fire Hazard: 1

  Reactivity: 0

  Personal Protection: E

National Fire Protection Association (U.S.A.):

  Health: 2

  Flammability: 1

  Reactivity: 0

  Specific hazard:

Figure 8.2 An MSDS file for benzhydrol. *(continued)*

Protective Equipment:
Gloves.
Lab coat.
Dust respirator. Be sure to use an
approved/certified respirator or
equivalent.
Splash goggles.

Section 16: Other Information

References: Not available.

Other Special Considerations: Not available.

Created: 10/09/2005 04:20 PM

Last Updated: 11/06/2008 12:00 PM

The information above is believed to be accurate and represents the best information currently available to us. However, we make no warranty of merchantability or any other warranty, express or implied, with respect to such information, and we assume no liability resulting from its use. Users should make their own investigations to determine the suitability of the information for their particular purposes. In no event shall ScienceLab.com be liable for any claims, losses, or damages of any third party or for lost profits or any special, indirect, incidental, consequential or exemplary damages, howsoever arising, even if ScienceLab.com has been advised of the possibility of such damages.

Figure 8.2   An MSDS file for benzhydrol. *(continued)*

Table 8.1   Unit change and material recovery.

Raw Material	CAS Number	Pounds
Substituted phenol	Proprietary	960.0
50% formaldehyde	Proprietary	243.9
Anti foam	Proprietary	0.1
Acid catalyst #1	Proprietary	2.58
Acid catalyst #1	Proprietary	1.96
Amine	Proprietary	2.46
Product Yield		1000.00

HEAT AND MASS BALANCE

Heat and mass balance for a chemical synthesis is illustrated in Table 8.2 with a block flow diagram in Figure 8.4. Due to proprietary nature, chemicals are not identified. The illustration suggests two steps. Any chemical engineer and/or a chemist should be very familiar with this heat and mass balance and can use this illustration for their process.

Health	2
Fire	1
Reactivity	0
Personal Protection	E

Material Safety Data Sheet
1,3-Diphenyl-2-thiourea MSDS

Section 1: Chemical Product and Company Identification

Product Name: 1,3-Diphenyl-2-thiourea

Catalog Codes: SLD2308

CAS#: 102-08-9

RTECS: FE1225000

TSCA: TSCA 8(b) inventory: 1,3-Diphenyl-2-thiourea

CI#: Not available.

Synonym: Diphenylthiourea; Sulfocarbanilide; sym-Diphenylthiourea

Chemical Name: Thiocarbanilide

Chemical Formula: C13-H12-N2-S

Contact Information:

Sciencelab.com, Inc.
14025 Smith Rd.
Houston, Texas 77396

US Sales: **1-800-901-7247**
International Sales: **1-281-441-4400**

Order Online: ScienceLab.com

CHEMTREC (24HR Emergency Telephone), call:
1-800-424-9300

International CHEMTREC, call: 1-703-527-3887

For non-emergency assistance, call: 1-281-441-4400

Section 2: Composition and Information on Ingredients

Composition:

Name	CAS #	% by Weight
{1,3-}Diphenyl-2-thiourea	102-08-9	100

Toxicological Data on Ingredients: 1,3-Diphenyl-2-thiourea: ORAL (LD50): Acute: 50 mg/kg [Rat].

Section 3: Hazards Identification

Potential Acute Health Effects:
Hazardous in case of ingestion, of inhalation. Slightly hazardous in case of skin contact (irritant), of eye contact (irritant). Severe over-exposure can result in death.

Potential Chronic Health Effects:
CARCINOGENIC EFFECTS: Not available.
MUTAGENIC EFFECTS: Not available.
TERATOGENIC EFFECTS: Not available.
DEVELOPMENTAL TOXICITY: Not available.
Repeated exposure to a highly toxic material may produce general deterioration of health by an accumulation in one or many human organs.

Section 4: First Aid Measures

Figure 8.3 An MSDS file for 1,3-diphenyl-2-thiourea.

Eye Contact:
Check for and remove any contact lenses. In case of contact, immediately flush eyes with plenty of water for at least 15 minutes. Get medical attention if irritation occurs.

Skin Contact: Wash with soap and water. Cover the irritated skin with an emollient. Get medical attention if irritation develops.

Serious Skin Contact: Not available.

Inhalation:
If inhaled, remove to fresh air. If not breathing, give artificial respiration. If breathing is difficult, give oxygen. Get medical attention.

Serious Inhalation:
Evacuate the victim to a safe area as soon as possible. Loosen tight clothing such as a collar, tie, belt or waistband. If breathing is difficult, administer oxygen. If the victim is not breathing, perform mouth-to-mouth resuscitation. Seek medical attention.

Ingestion:
If swallowed, do not induce vomiting unless directed to do so by medical personnel. Never give anything by mouth to an unconscious person. Loosen tight clothing such as a collar, tie, belt or waistband. Get medical attention immediately.

Serious Ingestion: Not available.

Section 5: Fire and Explosion Data

Flammability of the Product: May be combustible at high temperature.

Auto-Ignition Temperature: Not available.

Flash Points: Not available.

Flammable Limits: Not available.

Products of Combustion: These products are carbon oxides (CO, CO2), nitrogen oxides (NO, NO2...).

Fire Hazards in Presence of Various Substances:
Slightly flammable to flammable in presence of heat.
Non-flammable in presence of shocks.

Explosion Hazards in Presence of Various Substances:
Slightly explosive in presence of open flames and sparks.
Non-explosive in presence of shocks.

Fire Fighting Media and Instructions:
SMALL FIRE: Use DRY chemical powder.
LARGE FIRE: Use water spray, fog or foam. Do not use water jet.

Special Remarks on Fire Hazards: As with most organic solids, fire is possible at elevated temperatures COMBUSTIBLE.

Special Remarks on Explosion Hazards:
Fine dust dispersed in air in sufficient concentrations, and in the presences of an ignition source is a potential dust explosion hazard.

Section 6: Accidental Release Measures

Small Spill: Use appropriate tools to put the spilled solid in a convenient waste disposal container.

Large Spill:

Figure 8.3 An MSDS file for 1,3-diphenyl-2-thiourea. *(continued)*

Poisonous solid.
Stop leak if without risk. Do not get water inside container. Do not touch spilled material. Use water spray to reduce vapors. Prevent entry into sewers, basements or confined areas; dike if needed. Eliminate all ignition sources. Call for assistance on disposal.

Section 7: Handling and Storage

Precautions:
Keep locked up.. Keep away from heat. Keep away from sources of ignition. Empty containers pose a fire risk, evaporate the residue under a fume hood. Ground all equipment containing material. Do not ingest. Do not breathe dust. Wear suitable protective clothing. In case of insufficient ventilation, wear suitable respiratory equipment. If ingested, seek medical advice immediately and show the container or the label. Keep away from incompatibles such as oxidizing agents.

Storage: Keep container tightly closed. Keep container in a cool, well-ventilated area.

Section 8: Exposure Controls/Personal Protection

Engineering Controls:
Use process enclosures, local exhaust ventilation, or other engineering controls to keep airborne levels below recommended exposure limits. If user operations generate dust, fume or mist, use ventilation to keep exposure to airborne contaminants below the exposure limit.

Personal Protection: Safety glasses. Lab coat. Dust respirator. Be sure to use an approved/certified respirator or equivalent. Gloves.

Personal Protection in Case of a Large Spill:
Splash goggles. Full suit. Dust respirator. Boots. Gloves. A self contained breathing apparatus should be used to avoid inhalation of the product. Suggested protective clothing might not be sufficient; consult a specialist BEFORE handling this product.

Exposure Limits: Not available.

Section 9: Physical and Chemical Properties

Physical state and appearance: Solid. (Powdered solid.)

Odor: Odorless.

Taste: Not available.

Molecular Weight: 228.31 g/mole

Color: White.

pH (1% soln/water): Not applicable.

Boiling Point: Not available.

Melting Point: 152°C (305.6°F) - 155 C

Critical Temperature: Not available.

Specific Gravity: 1.32 (Water = 1)

Vapor Pressure: Not applicable.

Vapor Density: Not available.

Figure 8.3 An MSDS file for 1,3-diphenyl-2-thiourea. *(continued)*

Volatility: Not available.

Odor Threshold: Not available.

Water/Oil Dist. Coeff.: Not available.

Ionicity (in Water): Not available.

Dispersion Properties: See solubility in water, diethyl ether.

Solubility:
Easily soluble in diethyl ether.
Insoluble in cold water, hot water.
Very soluble in chloroform, olive oil, and alcohol.

Section 10: Stability and Reactivity Data

Stability: The product is stable.

Instability Temperature: Not available.

Conditions of Instability: Excess heat, dust generation, incompatible materials

Incompatibility with various substances: Reactive with oxidizing agents.

Corrosivity: Not available.

Special Remarks on Reactivity: Not available.

Special Remarks on Corrosivity: Not available.

Polymerization: Will not occur.

Section 11: Toxicological Information

Routes of Entry: Inhalation. Ingestion.

Toxicity to Animals: Acute oral toxicity (LD50): 50 mg/kg [Rat].

Chronic Effects on Humans: Not available.

Other Toxic Effects on Humans:
Hazardous in case of ingestion, of inhalation.
Slightly hazardous in case of skin contact (irritant).

Special Remarks on Toxicity to Animals: Not available.

Special Remarks on Chronic Effects on Humans: May cause adverse reproductive effects and birth defects (teratogenic) based on animal test data

Special Remarks on other Toxic Effects on Humans:
Acute Potential Health Effects:
Skin: May cause skin irritation.
Eyes: Dust may cause eye irritation.
Inhalation: May cause respiratory tract irritation.
Ingestion: Harmful if swallowed. May cause gastrointestinal tract irritation with diarrhea, and hypermotility. May affect behavior(ataxia, analgesia, convulsions)and respiration(cyanosis).

Figure 8.3 An MSDS file for 1,3-diphenyl-2-thiourea. *(continued)*

Section 12: Ecological Information

Ecotoxicity: Not available.

BOD5 and COD: Not available.

Products of Biodegradation:
Possibly hazardous short term degradation products are not likely. However, long term degradation products may arise.

Toxicity of the Products of Biodegradation: The products of degradation are less toxic than the product itself.

Special Remarks on the Products of Biodegradation: Not available.

Section 13: Disposal Considerations

Waste Disposal:
Waste must be disposed of in accordance with federal, state and local environmental control regulations.

Section 14: Transport Information

DOT Classification: CLASS 6.1: Poisonous material.

Identification: : Toxic Solid, Organic, n.o.s. (Thiocarbanilide) UNNA: 2811 PG: III

Special Provisions for Transport: Not available.

Section 15: Other Regulatory Information

Federal and State Regulations: TSCA 8(b) inventory: 1,3-Diphenyl-2-thiourea

Other Regulations:
OSHA: Hazardous by definition of Hazard Communication Standard (29 CFR 1910.1200).
EINECS: This product is on the European Inventory of Existing Commercial Chemical Substances.

Other Classifications:

WHMIS (Canada): CLASS D-1A: Material causing immediate and serious toxic effects (VERY TOXIC).

DSCL (EEC):
R25- Toxic if swallowed.
S24/25- Avoid contact with skin and eyes.
S26- In case of contact with eyes, rinse immediately with plenty of water and seek medical advice.
S36/37/39- Wear suitable protective clothing, gloves and eye/face protection.
S45- In case of accident or if you feel unwell, seek medical advice immediately (show the label where possible).
S46- If swallowed, seek medical advice immediately and show this container or label.

HMIS (U.S.A.):

Health Hazard: 2

Fire Hazard: 1

Figure 8.3 An MSDS file for 1,3-diphenyl-2-thiourea. *(continued)*

Reactivity: 0

Personal Protection: E

National Fire Protection Association (U.S.A.):

Health: 2

Flammability: 1

Reactivity: 0

Specific hazard:

Protective Equipment:
Gloves.
Lab coat.
Dust respirator. Be sure to use an
approved/certified respirator or
equivalent. Wear appropriate respirator
when ventilation is inadequate.
Safety glasses.

Section 16: Other Information

References: Not available.

Other Special Considerations: Not available.

Created: 10/09/2005 05:14 PM

Last Updated: 11/06/2008 12:00 PM

The information above is believed to be accurate and represents the best information currently available to us. However, we make no warranty of merchantability or any other warranty, express or implied, with respect to such information, and we assume no liability resulting from its use. Users should make their own investigations to determine the suitability of the information for their particular purposes. In no event shall ScienceLab.com be liable for any claims, losses, or damages of any third party or for lost profits or any special, indirect, incidental, consequential or exemplary damages, howsoever arising, even if ScienceLab.com has been advised of the possibility of such damages.

Figure 8.3 An MSDS file for 1,3-diphenyl-2-thiourea. *(continued)*

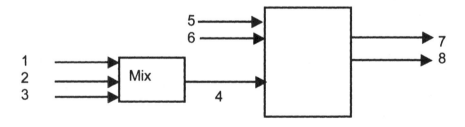

Figure 8.4 Process block flow diagram.

PROCESS CONDITIONS AND EFFECTS OF VARIABLES

This section is the preamble to the troubleshooting portion of the process. It concerns process-specific conditions and variables that need to be carefully monitored. For process efficiency and product quality, it is necessary that the defined process parameters be strictly followed. Process engineers will set process operating conditions. They need to be ob-

Table 8.2 Typical heat and mass balance.

Stream Number	1	2	3	4	5	6	7	8
Rate								
Pounds/batch	5000	2496	16325	23821	5360	1184	29945	420
Temperature, °C	50		25	47			85	85
Pressue, PSIG								
Physical Properties								
Specific gravity	1.232	1.525	1		1.334	1.525		
Viscosity, cps					1.7			
Specific heat, cp		0.783	1		0.38	0.783		0.617
pH								
Material								
Comp A	5000			226				
Comp B		1248				592	7	
Comp C		1248	16325	18135		592	18754	
Comp D					5360			
Comp E				5460			574	420
Comp F							4910	
Comp G							5700	

served. Since they, along with the manufacturing personnel, are most familiar with the process and the equipment, the causes and effects of raw-material variables and of equipment malfunction would be part of this section of the manual.

Process chemists and engineers might establish many of the operating parameters, values, and do's and don'ts, but the operating personnel have to live with minute-by-minute operation as they execute the process. Their input should be considered and included, as it does improve the overall process. For these reasons, the manual becomes a living document. For any batch or continuous process or for a blending operation, each operating parameter (temperature, pressure, concentration, flow rates, and any other variable) that can affect the product quality and yield should be defined.

Discussion clarifies in detail why the suggested process conditions are selected and the effects if the conditions are not followed. Deviation from the prescribed stoichiometry and temperature will result in less than optimum results. Operating parameters should be discussed. This section gives insight into the developers' thinking and knowledge base. It gives the readers a complete understanding of the process stoichiometry, temperature, and pressure variables. This information can be further used effectively to optimize the process.

Following is an example of this section. Names of some of the raw materials are excluded to protect confidentiality:

1. Initial raw-material charges must be accurate to ensure expected yield. The temperature conversion table must be used for raw material A to obtain correct charge. See page ∂ of this technical manual for raw material A charging instructions for the temperature density conversion table. At every charging of raw material A note its temperature and the volume charged. Inaccurate charge will lower product B yield.

2. The authorized charging and the reaction temperature of 47 +/- 2 degrees C must be maintained to ensure that the sodium salt of raw material A is in solution. The sodium salt of raw material A must be in solution to react with dimethyl sulfate.

3. Lab research has shown that the reaction gives the best yields at 47 +/- 2°C and at pH of 10.0. The pH should be maintained at 10.0 throughout the reaction of dimethyl sulfate with the sodium salt of raw material A.

4. The dimethyl sulfate addition rate has been preset. Any deviation from this preset rate without authorization of the process technology group should be reported immediately to the foreman.

5. At the end of the reaction, the pH is adjusted to pH 9.2–10.2 at 85°C to ensure that the unreacted raw material A is converted to its sodium salt. The sodium salt is water-soluble and would be removed from the product during the washing operation. If the pH is not adjusted to 9.8–10.2, unreacted raw material A will contaminate the product.

6. Hot water may be added to the reaction mixture to improve the phase separation. If the phase separation is slow, test by adding hot water to a sample. Hot water addition should improve the separation. Up to 500 gallons of hot water may be added to the batch but only after foreman's approval.

EQUIPMENT DESCRIPTION AND PROCESS AND INSTRUMENTATION DIAGRAM

In this section detailed equipment descriptions and P & I diagrams should be included. Chemical engineers are familiar with them. The following examples illustrate the concept. Control instruments and their specifications, heat exchangers and their sizes, and other equipment should be included. This document can be updated as equipment changes.

Tank 1:
Material of construction: 316 stainless steel
Dimensions: 107 inch height x 107 inch diameter
Capacity: 4100 gallons
Agitator
      Type: turbine
      HP: 10 HP
      RPM: 68
Coil: 2-inch schedule 10, 316 SS, 200 square feet
Pump 1:
Durco Mark III 1-1/2 × 1-6, 316 SS 1 HP XP
A P & I diagram (Figure 8.5) is included for illustration. Any person associated with process design would be very familiar with such diagrams.

SUGGESTED OPERATING PROCEDURE

Suggested operating procedures for a batch or continuous reactive chemical process and a batch or continuous blending process are translations and descriptions of a developer's process execution by a trained chemist or a chemical engineer with the help of manufacturing personnel. This has to be very carefully done, as changes to these procedures once the process has been approved internally or by any regulatory bodies that are involved in process approval becomes a challenge. I strongly believe that any changes in process operating methods should not be done until the change has been thoroughly tested and approved by developers and operations personnel. This is a good practice.

Often operating procedures become lengthy documents. In one instance the process had 10 steps and the product was isolated after each step for use in the next step. Each process step on an average occupied about five pages. Solvent recovery and reuse were separate procedures, resulting in a total for the operating procedure for one product of about 60 pages.

Procedures of such length not only become bulky to store but, in the case of a problem, it is not easy to review multiple batch records to pinpoint the problem. This also suggests that for a batch process there are too many steps and thus there is an opportunity to improve the process. A shutdown procedure should be part of the process. It is especially important for continuous processes. Safety considerations are an important part of the shutdown and startup practices.

LABORATORY SYNTHESIS PROCEDURE

Laboratory synthesis outlines what the developers were thinking when they were developing the product. The developer is sharing her/his rendition of the process and giving the "what and how" of the laboratory process. For blended products the procedures are fairly simple, as they outline the sequence of addition and whether any special in-process steps are needed. However, for chemical synthesis it is necessary that the laboratory synthesis be the one that is used in the commercial process.

Laboratory synthesis can be repeated by anyone to learn the developer's methods. It is also useful in the case of troubleshooting and optimization.

ADDITIONAL CRITICAL NOTES

This section should include cause-and-effect relationships that can be lead to potential unsafe situations. This is called HAZOP (hazard and operability) analysis. It provides solutions for each potential situation that can be recognized by the personnel who are involved in the design, scale-up, commercialization, and operation of the process. It is possible that not every situation will be covered, but every company has the expertise that can identify potential situations and how they can be prevented. Multiple books and articles have been written on the subject.[1, 2, 3, 4, 5, 6]

It is essential that every effort be made to ensure safe operation and safety of the personnel, assets, and environment.

ANALYTICAL METHODS

This section includes in-process analytical test methods and the expected results. The information should detail the tests, equipment, sampling techniques, and expected results if the process is operating as designed. It should also suggest desired frequency of sampling.

THERMODYNAMIC AND PHYSICAL PROPERTIES DATA

The physical property information of the raw materials and intermediates are of critical value, as it is used for the process design and for process finesse and manipulation.

Physical properties should include densities and temperature relationships, melting and boiling points, solubility data for the solvents being used, viscosity, surface tension, heat of reaction, specific heat, vapor pressure data, and any other physical property that will be useful in process development and design. Properties that are relevant to the process should be included in the manual.

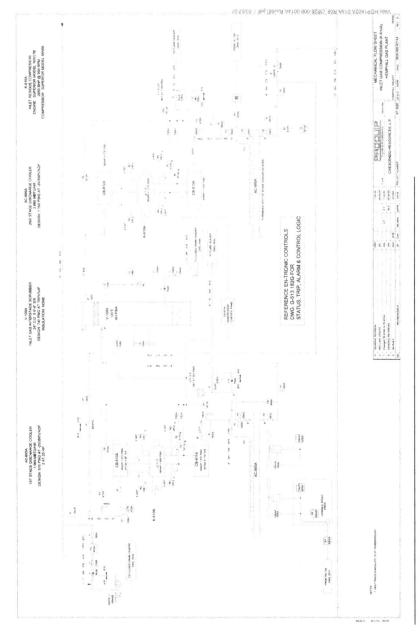

Figure 8.5 Typical P&I diagram.

At times this information is not readily available and might have to be generated or calculated using theoretical correlations. Table 8.3 includes some examples. Properties of raw materials and intermediates should be included. This might seem to be a significant challenge, but the benefits of the data collection last forever.

TROUBLESHOOTING GUIDE

This section of the manual is a living part of the document. Initially the developer, scale-up person, engineers, and manufacturing personnel add their methods and experiences to solve the problems that occur. After the initial startup, solutions to different operating problems are included, as they come into play.

PROCESS DESIGN MANUAL

The second critical manual is the process design manual for the process equipment. As I designed different processes, I documented the basis of the process equipment design. It was beneficial for troubleshooting, equipment replacement, and expansion.

It is a living manual. It elaborates the process equipment design basis and associated calculations. Individual companies can decide how and when they want to start on this manual. Preparation of the design basis starts when the process is close to being defined in the laboratory and is in transition to the pilot plant for scale-up.

Development of process design manual has to be done systematically. It starts with heat and mass balance, which is the first building block of the process. It defines the flow rates for each stream of the process, i.e., tells us the equipment size and its basis.

The next step is the preparation of process and instrumentation diagram. This defines how different the unit operations will execute in different unit processes. Process and instrumentation diagram is translation of operational vision of the process development chemist and/or chemical engineer to deliver quality product. It elaborates and ensures how the process operating conditions will be controlled per developing chemists/engineers vision.

Process equipment design for each equipment starts with the basic conditions i.e. flow rate, temperature and pressure. Pumps are designed to deliver the desired flow rate for smooth operation. Heat exchangers are designed to control the process thermal conditions for a safe process.

Table 8.3 Example of physical properties.

Heat of reaction can be calculated using the following formula: H_{rxn} = H_f (products) - H_f (reactants) H_{rxn} is heat of reaction, H_f heat of formation

Specific heat
    Acetic acid = 0.51 Btu/lb °F
    Water = 1.0 Btu/lb °F
    Methylene chloride = 0.289 Btu/lb °F

Viscosity
    Methylene chloride = 0.43 centipoise
    Tetrahydrofuran = 0.48 centipoise

Solubility gm/100 gm H_2O

Temperature °C	10	20	30	40
% NaOH	98	109	119	129

http://en.wikipedia.org/wiki/Solubility_table#S accessed July 23, 2009

Density   gm/ml
    Tetrahydrofuran = 0.886
    Methylene chloride = 1.326
    Dimethyl Formamide = 0.944
    Water = 1.0

Reactors/tanks are designed to proper reaction or blending operation. Pipe design's why, what and how are to be included. Process controls, their selection, sizing and their operating logic have to be defined and documented. All this is necessary for future for any training, trouble-shooting, equipment replacement and debottlenecking of the process. Documentation and compilation of paper work is the most challenging task but the most time saving task in case of operational upsets and malfunctions. Basically whatever equipment is on a Process & Instrumentation diagram, its design criterion and calculations have to be part of the Process Design Manual.

I used these two manuals extensively as my knowledge base to solve operational problems with different products. In addition, since the preparation of the suggested operating procedure was a cooperative effort between the operating and technical people, it produced the best operating procedure for a process.

These documents are considered by some to be mundane and not very useful for the operation. However, they are a compilation of the knowledge base and are useful later in process and product life.

REFERENCES

1. Center for Chemical Process Safety (CCPS), *Guidelines for Engineering Design for Process Safety*. John Wiley & Sons.
2. Center for Chemical Process Safety (CCPS), *Guidelines for Performing Effective Pre-Startup Safety Reviews*. John Wiley & Sons.
3. Center for Chemical Process Safety (CCPS), *Guidelines for Writing Effective Operating and Maintenance Procedures*. John Wiley & Sons
4. Center for Chemical Process Safety (CCPS), *Inherently Safer Chemical Processes: A Life Cycle Approach*. 2nd Edition. John Wiley & Sons.
5. Harris, R., Greenberg, Harris R., and Cramer, Joseph J. (editors), *Risk Assessment and Risk Management for the Chemical Process Industry*. John Wiley & Sons.
6. Center for Chemical Process Safety (CCPS), *Guidelines for Developing Quantitative Safety Risk Criteria*. John Wiley & Sons.

Chapter 9

Cross-Fertilization of Technologies

Industries use technologies and methodologies that seem industry-specific. However, for certain technologies the basic principles might be the same or similar. Over the years grass-roots technologies are improved, transformed, and take on an appearance that is different from their initial design or configuration. Since the basic premise of such technologies is the same, they should be considered and, if applicable, used more broadly if they can improve manufacturing processes or create new products.

Cross-fertilization of technologies might be considered an art, but it is not. It requires ability, creativity, and imagination to recognize and apply such methods, but it also takes significant internal selling to bring to fruition. Sometimes new ideas can be ahead of their time. Internal ideation, brainstorming, and selling are necessary and can be a challenge. Like everything we do, the biggest and the most prevalent hurdle in cross-cultural pollination of ideas is our own mindset. Saying "No" to an alternative and/or out-of-the-box concept is the simplest and the easiest response. It can permanently block consideration of a new idea. However, cross-fertilization can add significant value and create new opportunities for higher profits.

All of us are familiar with skunk works projects. They are a way to introduce and test new methods and technologies in business units without disrupting the daily routine. They are also a way to circumvent internal opposition to new developments and technologies. Skunk works have their advantages and disadvantages. This concept works, but its success is company-culture-dependent.

The steel industry used to be considered a routine, highly developed industry. Mini-mills changed the landscape. In the 1990s ArcelorMittal changed the business model completely and created a profitable business. Sometimes destructive paradigm shifts lead to innovation. The current global economic downturn will create new methods and leaders.

159

Arcade video games went through a revolution when Shigeru Miyamoto of Nintendo programmed an innovative chip from an air-bag system "accelerometer" to create Wii.[1]

Creative destruction and innovation are facilitated by cross-fertilization. In the mid-1980s no one would have thought that one day everyday print photographs would almost disappear and Kodak, an Americal/ global icon, would be in financial trouble. It happened due to cross-fertilization of electronic and image-capture technologies. In the automotive area Japanese companies led the way in quality innovation. This has resulted in the virtual disappearance of American automobile hegemony. Could the industries that have flourished but are not open to new ideas and innovation be considered "maladaptive?"[2]

REGULATORY CHALLENGE AND CREATIVE COMPETITION

Regulations can become a hurdle to innovation. If a company can deliver safe products and meet the regulatory challenges by incorporating technologies being used in other industries, the landscape for many businesses would be very different from what we have today.

In the early 1970s, the adoption of environmental regulations caused a seismic perturbation in global business. In every manufacturing segment many emission and discharge standards were adopted. Many companies innovated and thrived. However, due to a lack of desire for innovation, some companies in the developed countries closed or moved their operations to the developing countries.

Sometimes regulations, change, or a new competitor with a novel concept can nudge industries to innovate. However, sometimes companies are adamant and do not want to change. The North American automobile industry is an excellent example as it did not change and is almost on the verge of collapse.[3] Similarly, due to a lack of vision, the mobile-phone industry (excluding smart phones, i.e., iPhone or Blackberry) in the United States is behind the technologies being used elsewhere. I wonder if the pharmaceuticals are inching in that direction.

HOME-REPAIR PRODUCTS IN FOOD PACKAGING

In this environmentally conscious world, waste control is important. Most of us are familiar with the use of spackling compound to fill nail holes and cracks in our homes. Consumers purchase a pint or a larger size for repairs and paint a room or the whole house. After applying the spackling compound, we wait until it dries, then sand and paint. The leftover spackling compound is stored. It eventually dries up and is discarded as waste.

A few years ago, a laboratory developed a lightweight spackling compound that dried quickly and could be painted just after application. It did not require much sanding and did not shrink. Had it been packaged in a regular pint or quart can, it would have competed against other similar products on quality and price rather than performance. Sherwin-Williams decided to capitalize on performance, both reuse after an extended time and reduction of waste. I worked with my colleagues at Sherwin-Williams to develop a new package that would highlight the new attributes. Package design and graphics were developed for an impulse purchase.

The concept of the ketchup pouch exemplified convenience—just enough for a room, and an impulse purchase based on its newness, performance, design, and high profit margin. The biggest challenge was to ensure that the package contents did not dry, had a shelf life, and were convenient. Cross-fertilization of technologies was needed for success.

Sherwin-Williams approached a multi-layer barrier film producer that supplied the barrier film typically used for food packaging. A contract packager was engaged. They produced the pouches and filled them with the product. The package had an easy-tear nozzle opening for controlled liquid egress and volume dispensing. After use, the package could be rolled over and kept closed with a paper clip for use next time, even after six months. Reusability and reduced waste were ensured.

Attractive graphics showing use methodology (Figures 9.1, 9.2, 9.3, and 9.4) were used on individual packages as well as the box of 100. The box could be used for display and as a dispenser. Each individual package could also be pegged. The product was placed near the checkout register for impulse purchases, by other wall-repair supplies, and as part of a paint kit. The product was a commercial success with a very high profit margin. By using a moisture-barrier film used in food packaging, we successfully created a new product for the do-it-yourself (DIY) and contractor markets. Such a concept had never been used in the DIY market.

Innovation in packaging is continuously evolving and can be of particular value in product tamper proofing and preventing counterfeiting. This product was a result of cross-fertilization using a combination of technologies in barrier film, food, and coatings, as well as marketing methods, to capitalize on human impulse purchase behavior.

SPECIALTY CHEMICALS VS. ACTIVE PHARMACEUTICAL INGREDIENTS

Specialty fine chemicals, flavors, fragrances, resins, polymers, and other organic chemicals are synthesized to produce a desired quality product.

Figure 9.1   Box design left face. *(Used with permission.)*

Figure 9.2 Box design right face. *(Used with permission.)*

Figure 9.3   Box design front face. *(Used with permission.)*

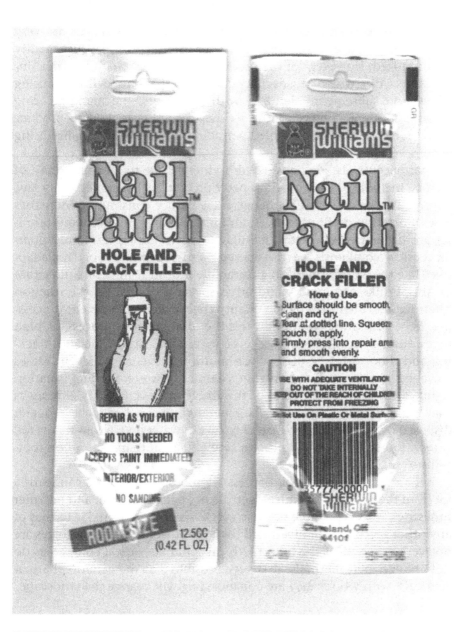

Figure 9.4   Pouch graphics on front and back panels.
(Used with permission.)

The current manufacturing technologies for these chemicals are very advanced. In-process sampling to test for intermediate product quality is minimized. However, for the production of active pharmaceutical ingredients (API), which are specialty chemicals that have a disease-curing value, and subsequent blending with excipients, the manufacturing technologies are not even close to the technologies used to produce their brethren. They are sampled periodically to check for quality using expensive equipment.

Pharmaceuticals[4] are considered invincible due to a combination of factors, including intellectual property protection, consumer safety law, and regulatory approval challenges. Regulatory constraints, industry's ease of meeting profit objectives, and internal apprehension and mindset are hindrances to better manufacturing technologies. Could pharmaceutical companies be considered "anti-panarchial"[1] or "maladaptive" like the North American automobile industry? In the manufacture of APIs, unlike commodity and other specialty chemicals, the reaction intermediates are isolated for a quality check before subsequent processing. Repeated sampling complicates the total business process, as high investment is needed for storage of inventories and associated tracking and monitoring. Processes use multiple solvents, which increases recycling needs and creates higher volumes of waste. Product isolation significantly reduces the yield of the whole process. Lower yield means that higher amounts of waste is generated, and it has to be properly disposed of to minimize release of toxins in water and land. The overall carbon footprint of an API process is much larger than the carbon footprint of similar chemicals.

API and drug-formulation process technology developers can learn a lot from the specialty chemical, petroleum, petrochemical, and polymer industries. This is possible because synthesis of APIs is similar to that of any other specialty chemicals and their manufacturing technologies are not very different. If the brand (ethical) pharmaceutical companies do not change their methods, it is very possible that the generic pharmaceuticals[5] with whom they are collaborating will change the landscape.

LIQUID BLENDING PROCESSES

Many products are produced by dispersing solids in liquids to which other liquids are added to achieve a uniform blend for the desired applications. Generally these products fit the broad classification of coatings that enhance the visual effect, such as architectural, automotive, industrial, paper, printing, and cosmetic (lipsticks and nail polishes).

These are different applications, but the basic theme of solid-liquid dispersion is the same. Solids are wetted, reduced to appropriate size, and admixed with necessary ingredients to produce a product of desired performance.

Can the commercially available equipment being used to produce coating X be used to create a better process for coating Y? My conjecture is that it is possible. We have to review such possibilities. Liquid coatings are produced using batch processes. It is possible that a staid batch process could be converted to a continuous process.

Rotor/stator-type equipment has been used in reducing the particle size. It should be explored in pharmaceuticals to get a uniform size during crystallization[6] of API.

Practices of specialty chemicals and liquid blending can be applied to coatings, resins, and polymers. One has to review the opportunities. Many times they are right in front of us.

In the manufacture of coatings, the use of heat exchangers is generally not considered. However, with changing environmental laws, the use of heat exchangers that have primarily been used in chemical manufacturing operations can be very beneficial, especially in the manufacture of solvent-based coatings. They can lower solvent emissions and can also reduce process cycle time, thereby improving process productivity and profitability. An example of using a "home-made" heat exchanger is discussed in Chapter 2.

LABELING

Container labeling is considered a simple technology of delivering a self-adhesive label to a container as it passes through. Battery labels are self-adhesive shrink labels. In the 1990s battery-life indicator devices[7, 8] were introduced. They are a combination of electrical engineering and print technology applied to a shrink film. It is a complex construction created by cross-fertilizing different physical sciences.

HOT-MELT EXTRUSIONS

Hot-melt extrusion technology is used in the manufacture of plastics and adhesives. It is also being explored in pharmaceuticals.[9] It is necessary to exploit and explore technologies when they can improve and facilitate product usage.

Spray dried water- based coating powder can be used as a filler in plastic extrusions (Chapter 3). In an environmentally conscious world, this low cost powder can be of value. If we do not look at such alternates, it will become landfill and no one knows its long-term impact on soil.

MINIATURIZATION/SIZE REDUCTION

With the advent of improved electronic and computer technologies, miniaturization is being used to our advantage in many areas. Ultrasound used to require big and stationary equipment. Now ultrasound equipment has been miniaturized and is available as portable equipment. Such equipment has been of tremendous value for people with minimum healthcare facilities.

Prior to the 1980s, computers were huge in size and needed large rooms and cooling facilities. We never thought of laptops and hand-held smart devices like the Blackberry, iPhone, iPad, Kindle, etc. becoming part of our everyday lives. I would not be surprised if these technologies become outdated in the next five to ten years.

Commercially, chemicals are produced using reactors, heat exchangers, and other associated equipment that is generally large in size. Due to limitations in fabrication technology and need, chemical equipment has not been miniaturized or reduced in size. Microreactors (Chapter 6) are still a laboratory curiosity. They could be good for the manufacture of certain fine/specialty chemicals and active pharmaceutical ingredients (API) if an effort were made. There are limited numbers of suppliers for the microreactors and they are expensive, limiting their use. If properly developed and incorporated, they would have a large impact on reactive chemical manufacturing. They even could make the handling and use of hazardous chemicals safe. Currently, very expensive methods and technologies are used for personnel safety.

Creative use of commercial and laboratory size equipment could be used for the manufacture of chemicals, but has not been considered since it is not normal for the chemical manufacturing operations. This could lead to the development of process technologies that could easily move batch processing to continuous processes. It could reduce solvent use and improve process yields, thereby lowering the carbon footprint. Creativity and imagination is paramount.

ALTERNATE ENERGY

With the recent turmoil in petroleum supplies we all saw petroleum and derivative prices fluctuate all over the place. Alternate energy sources have come on the table more quickly and often. Human creativity and ingenuity will create opportunities. Significant work is being done. A recent report[10] describes one of many alternates being reviewed. Many will involve technologies that have been developed in the past and would be improved to deliver economical fuel for growing needs.

REFERENCES

1. Ramo, Joshua Cooper. *The Age of The Unthinkable*. Little, Brown and Company 2009;118–126.
2. Hollings, C. S. "Understanding the Complexity of Economic, Ecological and Social Systems." Springer-Verlag, *Ecosystems*. 2001;4; 390–405.
3. Ingrassia, Paul. *Crash Course: The American Automobile Industry's Road from Glory to Disaster*. Random House. January 2010.
4. Malhotra, Girish. "Pharmaceutical Costs, Technology Innovation, Opportunities & Reality." *Pharmaceutical Processing*. February 2010.
5. Acharya, Satish. The Productivity Tiger—Time and Cost Benefits of Clinical Drug Development in India. http://pharmalicensing.com. Accessed November 18, 2009.
6. Docherty, Robert, Kougoulos, T., and Horpool K. Materials Science and Crystallization: The Interface of Drug Substance and Drug Product. *American Pharmaceutical Review* Vol. 12. 6;September/October 2009;34–43.
7. Bailey, John C. Temperature-Compensated Thermochromic Battery Tester. USP 5,841,285. November 24, 1998.
8. Weiss, Victor H., et al. Battery Tester Label for Battery. USP 6,054,234. April 25, 2000.
9. Repka, Michael A. "Hot Melt Extrusion." *American Pharmaceutical Review*. Vol. 12. 6;September/October 2009;18–27.
10. Beyond Biofuels—Carbon-neutral WindFuels.™ Doty Scientific, Inc. Clemson, SC.

Chapter 10

Scale-up to Commercialization

Commercialization is scaling up a laboratory product into a marketable product using a manufacturing process that is safe, economical, repeatable, and environmentally sustainable and that produces a quality product the first time and all the time. It is the most exhilarating, challenging, and grueling process and experience. The product could be derived from a chemical reaction or a blend of chemicals.

Invariably the process is scaled up in a pilot plant. Experiences in laboratory, pilot plant, and commercial-scale operation are a source of continuous learning and improvement. These experiences, when integrated, will deliver an economical process.

There are many ways to scale up and commercialize products developed in laboratories. I have outlined the methods that have helped me to successfully scale up processes. Readers can modify these and/or develop their own methods. Books and articles[1, 2] have been written on this subject and can be reviewed to select the method that best suits the readers. Most of these cover various unit operations. My focus is on chemical-reaction and liquid-blending scale-ups and some of the associated unit operations.

Laboratory development processes show us a path to produce a product either through the reaction of different chemicals or through the blending of different chemicals. For products that are based on blending of chemicals, their formulation can be nearly optimized in the laboratory, as the products are generally developed to deliver a certain performance. These formulations are then scaled up to produce products on a large scale. Commercial-scale equipment might require some optimization. This effort can be minimal.

However, the effort with the reactive process has to be much higher. This is due to differences in the laboratory equipment capabilities vs. what can be achieved on a larger scale-up using larger commercial equipment. Laboratory processes are generally dilute, as they are just showing the feasibility of the process and not the actual commercial process. Scale-up and commercialization is the prime location to develop a process that will result in the highest yield and reduce the amount and the number of solvents used in the reactive processes. This effort reduces the infrastructure costs for solvent storage and the costs associated with their recovery. It reduces product manufacturing costs. It also reduces the waste-treatment effort and related costs. Environmental impact can be significantly reduced.

A review of the following helps me in quickly learning the processes and allows me to facilitate scale-up and commercialization. This review/ checklist allows one to consider alternative processing options and methods that are simple, safe, economical, innovative, and environmentally sustainable:

- Chemistry and development process
- Mass and heat balance
- Stoichiometry
- Physical properties of the chemicals involved
- Observation of the reaction and/or blending process in the laboratory
- Pilot-plant process equipment.

Every effort must be made to reduce the number of unit operations and solvents in each process and process step. Processes should be operated at the highest feasible concentration. Such an effort reduces batch cycle time, capital investment, and process waste. These considerations should be part of the process review.

The above effort can also suggest improvements and finessing of the process and chemicals. Any process changes based on the review should be discussed with the chemistry developers to determine their safety and feasibility. These changes and any process alternatives should be tested in the laboratory before any large batch is planned. If there are gaps and additional information such as reaction mechanisms or physical properties is needed to assist in the scale-up, it should be assembled and/or generated.

REACTION SCALE-UP/COMMERCIALIZATION

Preparation of 6-chlor-5-(2-chloroethyl) oxindole[3, 4] is used as an example. All of the observations and comments are my own and are in no way a reflection on the developers' capabilities. I have applied the methods outlined above to lay out my scale-up plan.

Process Outline

Following is the synthesis.[3] A dry 50 L reactor was charged with methylene chloride (10 L) and anhydrous aluminum chloride (6.56 kg, 49.2 mol, 3.3 equiv) under nitrogen. The contents were cooled to 10–15°C and stirred for 15 minutes. Chloroacetyl chloride (2.70 kg, 23.9 mol, 1.6 equiv) was added over two hours. To this reaction mixture was added 6-chlorooxindole (2.50 kg, 14.9 mol, 1 equiv) as a solid. The reaction mixture was stirred and heated at 30–40°C under nitrogen atmosphere until an in-process control sample indicated completion of reaction by HPLC assay. At the end of the reaction period, the mixture was cooled to 0–5°C. To the reaction mixture was slowly added 1,1,3,3-tetramethyl-disiloxane (TMDS) (4.01 kg, 29.8 mol, 2.0 equiv). The reaction mixture was stirred at this temperature for four to six hours. After completion of the reaction as judged by the HPLC assay, the reaction mixture was quenched with the addition of water (30–40 L) over three hours. A caustic scrubber scrubbed the HCl gas that evolved during this quench. The reaction mixture was distilled at atmospheric pressure until the temperature reached 50°C to remove most of the methylene chloride. The reaction mixture was cooled to 25°C. Tetrahydrofuran (65.7 L) was charged, and the reaction mixture was agitated and heated to dissolve all solids. The reaction mixture was allowed to settle. The lower aqueous phase was separated and discarded. The organic layer was concentrated to a volume of 10–15 L by distillation under reduced pressure. The resulting product slurry was cooled to 25°C, and isopropanol (5 L) was added to the slurry. The reactor contents were cooled to 0–5°C and stirred for two hours. The product slurry was filtered through a large (I.D. 457 mm) Buchner filter funnel. The product was washed with isopropanol (1 L). The isolated material was dried in a tray drier at 40–50°C for 12 hours to yield 2.98 kg (86.8 percent yield) of the desired product 6-chlor-5- (2-chloroethyl) oxindole.

Laboratory Process Review

European and U.S. patents are referenced[3] for the products and intermediates. The reason for the development of the above process is the toxicity of 5-chloroacetyl 6 chloro 2-oxindole. It is a strong skin sensitizer. One pot process eliminates the isolation of this intermediate. An alternative method[4] has been suggested to bypass 5-chloroacetyl 6 chlo-

Table 10.1   Effect of number of steps on yield.

Yield per step	95	80	70
Steps	Overall yield %		
5	77.4	32.7	16.8
10	59.9	10.7	2.8

ro 2-oxindole. However, due to multiple steps the process yield drops significantly. Table 10.1 illustrates the effect of combining a number of steps and their yield. The ideal process should have a minimum number of process steps, and the intermediates should not be isolated. For batch processes, such an effort will also reduce the cycle time, improve yield, and reduce costs.

Generally in the manufacture of specialty chemicals that have disease-curing value (APIs), intermediates are isolated and tested to ensure quality. Such has been the tradition in the manufacture of pharmaceuticals. If we have a complete understanding and control of the chemistry, we would not need to isolate the intermediates. For example, 6-chlor-5-(2-chloroethyl) oxindole is a precursor for ziprasidone hydrochloride and, based on my review, it is not necessary to isolate any intermediates. To get a better understanding of the process chemistry, it is helpful to break down the reaction into its steps. This also allows a better understanding of the process steps.

The chemical reaction is illustrated in Figure 10.1.[5] Based on the above process description and the reaction scheme of Figure 10.1, 6-chlorooxindole reacts with chloroacetyl chloride in presence of aluminum chloride to produce CP-89,574 (5-chloroacetyl-6-chloro-2-oxindole), which in turn is converted to 6-chlor-5-(2-chloroethyl) oxindole, in the presence of tetramethyldisiloxane (TMDS) and aluminum chloride. Each

Table 10.2   Stoichiometry comparison.

Chemical	Moles	Mole ratio	
		Experiment	Theoretical
6 chloro oxindole	14.9	1.0	1.0
Chloroacetyl chloride	23.9	1.6	1.0
AlCl$_3$	49.2	3.3	3.0
Tetramethyldisiloxane (TMDS)	29.8	2.0	2.0

Table 10.3 Physical properties of raw materials and intermediates.

Chemical	Methylene chloride	6 chloro oxindole	Chloroacetyl chloride	AlCl$_3$	TMDS	5-(2-chloroacetyl)-6-chloro-2-oxindole	6-Chloro-5- (2-chloroethyl) oxindole
CAS #	75-09-2	56341-37-8	79-04-9	7446-70-0	3277-26-7	118307-04-3	118289-55-7
MW	85	167.6	113	133	134.2	244	230
MP, °C		195-196		192.4		202-206	226-228
BP, °C	40		105-106		70-71		

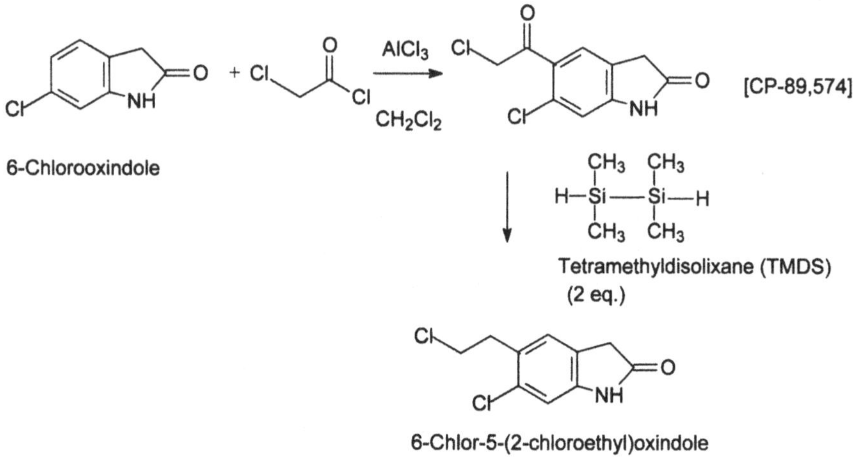

Figure 10.1 Chemistry of 6-chlor-5-(2-chloroethyl)oxindole.

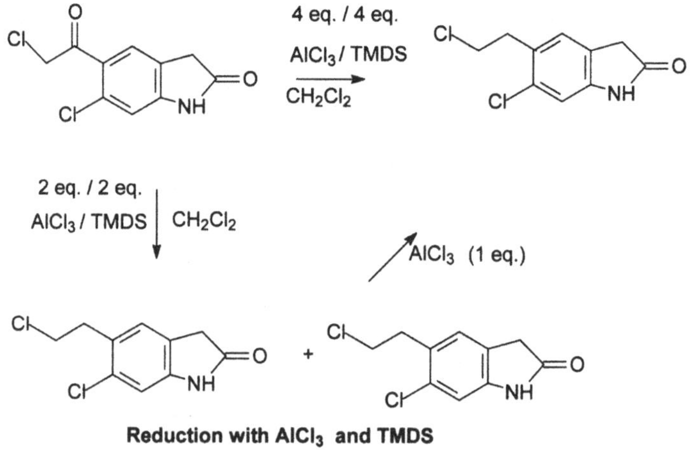

Figure 10.2 Alternate # 1 chemistry of 6-chlor-5-(2-chloroethyl)oxindole.

step is a simple Friedel-Crafts reaction. A possible reaction mechanism is explained in Figures 10.2 and 10.3. Stoichiometry suggested in the above experiments is compared with the theoretical stoichiometry in Table 10.2. The physical properties of some of the raw materials and intermediates are tabulated in Table 10.3.

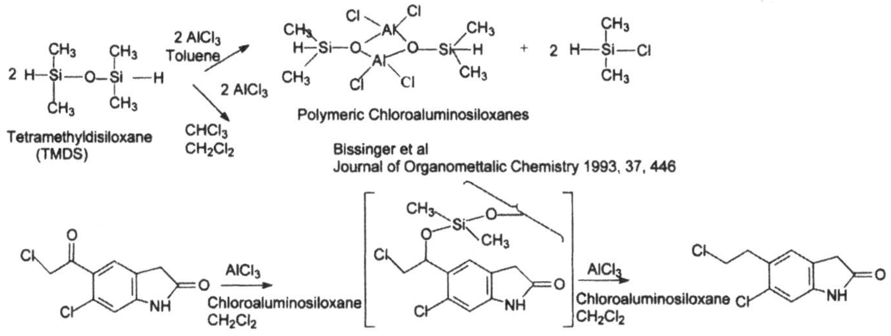

Figure 10.3 Alternate # 2 chemistry of 6-chlor-5-(2-chloroethyl)oxindole.

A careful review suggests that about 47 percent (weight for weight) aluminum-chloride slurry in methylene chloride is produced. About 2.7 kilograms of chloroacetyl chloride is added over two hours to this slurry. This is suggestive of an exotherm.[6] We need to understand this exotherm prior to the scale-up. We need to know why it happens. Is there a chemical reaction, and if so how it can be controlled without any adverse effect on the process? Hydrochloric acid gas is generated during the reaction with 6-chlorooxindole and must be scrubbed with caustic. The product 5-(2-chloroacetyl)-6-chloro-2-oxindole is a solid. We have slurry in methylene chloride. The process should not need any nitrogen purge, as it is being conducted at and around the boiling point of methylene chloride.

In the second reaction step, TMDS is added at low temperature, and the reaction mass is stirred for four to six hours to produce 6-chloro-5-(2-chloroethyl) oxindole. Methylene chloride is removed by distillation, and tetrahydrofuran is added. The reaction mass is washed. The reaction mass is concentrated, and 6-chloro-5-(2-chloroethyl) oxindole is crystallized using isopropanol.

Based on my analysis of the chemistry, the laboratory batch process can be significantly simplified. There is nothing wrong with the current method outlined in the referenced paper. The suggested process is what a chemist would follow in a laboratory. For a commercial process, I would like to consider alternative processing methods and compare them to the suggested laboratory process. The most economically viable and safe process should be commercialized.

Alternative Process Stoichiometry

Comparison of the experimental and theoretical stoichiometry suggests that 60 percent excess chloroacetyl acetyl chloride is being used in the process. This presents an opportunity. If we can reduce the chloride use, it will reduce the batch cycle time by reducing addition and completion time of the reaction step. In addition, we will have less waste to treat, as the unreacted chloride needs to be neutralized and handled in the waste treatment. Conservation should always be considered for every process step. If the amount of chloroacetyl chloride were reduced, the total slurry concentration would need a review, as acetyl chloride is a liquid. It might become necessary to increase the amount of methylene chloride to have a mixable slurry.

The process suggests that 3.3 moles of $AlCl_3$ are needed. There is 10 percent excess. This is not very high, but can be optimized for a commercial process after a scale-up batch has been produced. Following are a few of the alternate processing options that should be considered for a scale-up batch using the existing laboratory process:

Consider using an alternative nonreactive solvent that will dissolve the solids. The rationale for such a solvent is to improve the mass and heat transfer. The solvent should have a low temperature azeotrope with water, as this property can be used to remove the solvent after the second reaction step if necessary. If the solids are soluble, it could eliminate the filtration and simplify purification and crystallization.

Speed up the addition rate of chloroacetyl chloride and control the generated exotherm using chilled water in the reactor jacket. This should reduce the addition time that is needed in the laboratory. Better heat-transfer equipment will assist the process. As indicated earlier, it is necessary to understand the reasons for this exotherm.

If the alternative solvent is a higher boiler than methylene chloride, raising the reaction temperature, as 6-chlorooxindole is added, should be considered. A higher temperature will improve the reaction rate. If the solvent evaporates, its evaporation, condensation, and recycling can be used to control the exotherm.

TMDS addition rate could be increased to speed the reaction rate. The reaction mass temperature should be controlled at a temperature to maximize conversion and minimize reaction time. Solvent evaporation, condensation, and recycling can be used to control the exotherm and reduce the step reaction time.

The process calls for dissolution in tetrahydrofuran, phase separation, concentration of the THF layer, and crystallization using isopropanol. Yield is about 87 percent. Three solvents (methylene chloride, THF, and

IPA) are used in the laboratory process. If 6-chloro-5-(2-chloroethyl) oxindole is not isolated, it will eliminate the use of THF and IPA. 6-chloro-5-(2-chloroethyl) oxindole, after methylene chloride removal, can be converted in the next step to produce ziprasidone hydrochloride. This should improve the overall yield of the ziprasidone process.

Safety of the overall process on a commercial scale increases if we do not isolate any intermediates. In addition, one need not worry about the toxicity of chemicals and intermediates. Since the feed rates and their method of addition will change in comparison to the laboratory process, their impact on conversion and intermediate purity needs to be tested.

Other Process Options

Other processing options are worth exploring. In the alternative process, it is necessary to reduce the molar excess chloroacetyl chloride. All of the aluminum chloride might not be necessary in the first step. About one mole of $AlCl_3$ is needed in the first step. The remainder of the aluminum chloride can be reacted with TMDS to produce polymeric chloroaluminosiloxanes, which would be added to the reaction mass from the first step. This addition method will confirm the postulated reaction mechanism and improve control of the process. This alternative should allow minimal excess of reactants. It could improve the stoichiometry, process conditions, and yield and reduce the overall batch cycle time.

Yield and the processing conditions for the alternative process would have to be reviewed and confirmed. This could also validate use of one solvent with water instead of three solvents by making the process a safe (no intermediate isolation of hazardous materials) and environmentally sustainable process.

The above alternative scheme might look like it would expand the process in multiple reactors, but that is not the case. It looks at how the process chemistry can be simplified, conducting it in a way that the process produces quality without intermediate isolation and has high throughput. It is schematically shown in Figure 10.4. Information developed will be used to improve the batch process. This information could also facilitate the development of a continuous process.

Potential Continuous Process

If the above alternative process gives the anticipated results, there is an excellent possibility to produce ziprasidone hydrochloride using a continuous process. A potential block diagram for a continuous process is illustrated in Figure 10.4. The concept outlined would have to be tested.

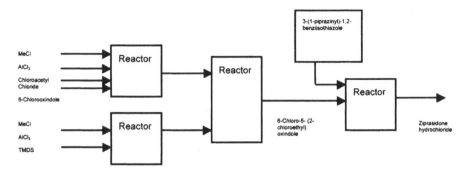

Figure 10.4   Potential continuous process for Ziprasidone hydrochloride.

The data developed from the laboratory synthesis and the alternative outlined above would be very helpful.

The following continuous process for ziprasidone hydrochloride[7, 8] is proposed. We should not forget that this is a specialty chemical that happens to cure a disease. The process must be safe, repeatable at all times, with quality tightly controlled within specifications. Rework and isolation of intermediates are not an option.

In a continuous process, chloroacetyl chloride is reacted with 6-chlorooxindole in the presence of aluminum chloride to produce chloroacetyl chlorooxindole. Chlorooxindole is converted to chloroethyl oxindole, which is then reacted with 3-(1-piperazinyl)-1,2-benzisothiazole to produce ziprasidone hydrochloride. It is purified through crystallization. Since there is no intermediate isolation, one solvent in addition to water may be necessary to produce the final product. In the selection of the solvent, solubility of the raw materials and intermediates along with other physical properties and cost are considerations.

A process flow diagram for the continuous process is suggested in Figure 10.5. Tank #1 is an aluminum chloride and solvent slurry tank. Provisions must be made to keep the aluminum chloride suspended uniformly. There are methods to achieve this suspension. Its flow to reactor #3 must be controlled in ratio with the slurry of chloroacetyl chloride and 6-chlorooxindole. Tank #2 is used to make slurry from the acetyl chloride and oxindole. It may be necessary to add some solvent (the same solvent used in the aluminum-chloride slurry) to keep this slurry uniform, as it is fed to reactor #1 for reaction with aluminum chloride. We must be mindful of this reaction, as the laboratory process has an exotherm when acetyl chloride is added to the aluminum-chloride

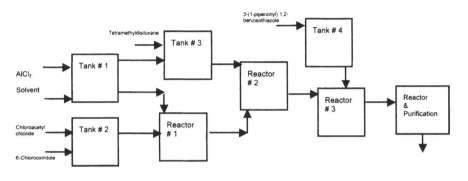

Figure 10.5 Alternate continuous process for Ziprasidone hydrochloride.

slurry. In the laboratory process, methylene chloride is removed by distillation. Because there will be no isolation of any intermediate, it may
not be necessary to remove the solvent. Thus, a solvent that dissolves
various solids and can be used for crystallization can be very beneficial
to the whole process.

Slurries from tanks #1 and #2 will be ratio-controlled to reactor #1,
which must have sufficient residence time to produce 5-chloroacetyl-
6-chloro-2-oxindole. Slurry of oxindole is fed to reactor #2, which reacts with the polymeric chloroaluminosiloxanes produced in tank #3.
Reactor #2 must have sufficient residence time to produce 6-chloro-5-
(2-chloroethyl) oxindole. Tank #4 contains benzisothiozole slurry. This
slurry is fed to reactor #3. Reactor #3 and its associated equipment complete the conversion of oxindole to ziprasidone hydrochloride, and it is
processed to produce the desired product meeting established specifications. Any chemist/chemical engineer well versed in the scale-up of the
specialty chemicals will understand the process.

COMMERCIALIZATION OF FLUOROCHEMICAL PRODUCTS

PolyFox™ fluorochemicals are a family of products available from OM-
NOVA Solutions. When incorporated in various coatings and applications, they enhance their functionality. PolyFox™ polymers are produced from their respective monomers. Polymers are converted to other
downstream products. Various patents[9, 10, 11] describe the synthesis of
these products. Chemistries[12, 13] are outlined in Figure 10.6.

In the scale-up of these products, the six steps outlined at the begin-

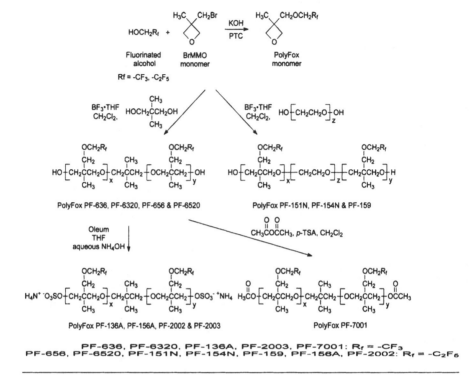

Figure 10.6 PolyFox chemistries.

ning of the chapter were followed. The scale-up of the family of products was divided into monomer, polymer, and their resultant products.

Monomer Laboratory Preparation

This process describes the monomers (3-methyl-3-((2,2,2-trifluoroethoxy) methyl) oxetane, referred to as 1a, and 3-methyl-3- (2,2,3,3,3-pentafluoropropoxy)methyl) oxetane, referred to as 1b.

A three-necked, 10 L jacketed vessel with heater/chiller bath, thermometer, stir bar, condenser, addition funnel, and inert-gas inlet and outlet was preheated to 85°C. An aqueous solution (12.2 wt %, 1.97 kg, 0.751 mol) of the phase-transfer catalyst, tetrabutylammonium bromide (TBAB), was added to the reaction vessel and allowed to stir until the catalyst was dissolved. Then 3-bromomethyl-3-methyloxetane (5.04 kg, 30.6 mol) and 2,2,2-trifluoroethanol (3.0 kg, 30.0 mol) were added to the vessel and purged with nitrogen. The addition funnel was charged with a 45 percent aqueous solution of KOH (4.11 kg, 32.967 mol). The KOH solution was added to the vessel as quickly as possible without allowing the reaction exotherm to increase the solution temperature above 100°C. The reaction was allowed to proceed for four hours. Water (30 g) was added when the solution temperature cooled to 20 to 60°C. A blue indicator can be added to the bromophenol to enhance the visibility of the phase split. The two phases separate in approximately 2 hours, after which time the aqueous layer was separated and discarded. A total of 5.84 kg of crude monomer was isolated. Heptane (600 g) was added to the reaction mass, and the organic phase was isolated by azeotropic vacuum distillation (82 to 86°C, ~ 28 mmHg). The product was isolated as a clear, low-viscosity liquid in 90 percent yield. The purity of monomer 1a was determined by ^{1}H NMR spectroscopy. Monomer 1b was prepared in identical fashion using 2,2,3,3,3-pentafluoropropanol at 90 percent yield. Polymerization of monomers 1a and 1b is sensitive to water. Water content was determined by Karl Fisher titration. If the water level is too high (greater than 0.20 percent), the monomer can be distilled at 82–86°C, ~ 28 mmHg.

Monomer Scale-Up

Stoichiometry of the laboratory process suggests that the minimal excess of raw material is being used. It should be optimized further, as any reduction results in savings. Since all of the additions and product are liquid, the process becomes very simple. A review of the experimental procedure suggests that control of the exotherm can reduce the KOH addition time. After the reaction is completed, it is quenched with water. The amount of quench water can be controlled so that potassium bromide stays in solution and minimum aqueous phase is incorporated in the organic phase. The density difference of the phases can be controlled. The phases can be easily separated using a gravity decanter (discussed in Chapter 4). A high differential between densities will accelerate the phase separation, and it should be a clean separation. The laboratory procedure suggests azeotroping the water with heptane. Since the phase separation would be very clean, use of heptane can be eliminated on a commercial scale. The monomer can be purified by distillation.

Polymer Laboratory Preparation

This process describes the polymers [α,ω-(dihydroxy) poly(3-methyl-3-[(2,2,2-trifluoroethoxy)methyl] oxetane), referred to as 2a, and [α,ω-(dihydroxy) poly(3-methyl-3-[(2,2,3,3,3-pentafluoropropoxy)methyl] oxetane), referred to as 2b.

A three-necked, 10 L jacketed vessel with heater/chiller bath, thermometer, stir bar, condenser, addition funnel, and inert-gas inlet and outlet was charged with dried neopentyl glycol (329.92 g, 3.17 mol), $BF_3 \cdot THF$ (177.26 g, 1.27 mol) catalyst, and CH_2Cl_2 (1.86 kg, 21.84 mol) solvent cooled at 25 to 30°C. The neopentyl glycol was dried by dissolution in toluene, and the solvent was removed under reduced pressure. The initiator and catalyst solution was allowed to stir for 30 minutes at room temperature under a positive pressure nitrogen purge. Monomer 1a (3.50 kg, 19.01 mol) was then added to the catalyst/initiator solution at a rate of 50 g/minute using a pump while the reaction temperature was maintained at 35 ± 10°C. The reaction was allowed to stir for two hours. Extra CH_2Cl_2 was added (2.8 kg, 32.97 mol). Residual $BF_3 \cdot THF$ was removed by washing with 2.5 weight percent sodium bicarbonate and a water rinse at 40°C. Solvent was then removed under reduced pressure at 80°C. Polymer 2a was obtained as a clear, viscous liquid in 95+ percent yield. The degree of polymerization was determined using ^1H NMR spectroscopic end group analysis and found to be 7 ± 0.1. Polymer 2b was prepared similarly in 95+ percent yield.

Polymer Scale-Up

Based on the above description and applying the guidelines discussed earlier, the process is easy to scale up. Addition rate of the monomer would be dependent on the available cooling capacity. The initial slurry is about 14 percent glycol in methylene chloride. The amount of methylene chloride can be reduced—i.e., the glycol concentration increased. This is possible because the monomer is a liquid, making the reaction an all-liquid system. Batch productivity will be improved with methylene-chloride reduction. Viscosity data would assist the scale-up. An all-solution process will have good heat transfer and good reaction rate. Equipment configuration and capabilities will dictate the dilution needs. Neutralization and phase separation would be easy due to density differences. The polymer would be dried using conventional process equipment. Removal of methylene chloride would require conventional high vacuum equipment with necessary provisions to ensure that methylene chloride is not emitted in the air.

Polymer Derivatives Laboratory Preparation

This process describes the polymer derivatives [α,ω-(diammonium disulfato)poly(3-methyl-3-[(2,2,2-trifluoroethoxy)methyl]oxetane), referred to as 4a, and [α,ω-(diammonium disulfato)poly(3-methyl-3-[(2,2,3,3,3-pentafluoropropoxy)methyl]oxetane), referred to as 4b.

A 10 L jacketed reactor with heating and cooling capability was charged with [α,ω-(dihydroxy) poly(3-methyl-3-[(2,2,2-trifluoroethoxy) methyl]oxetane) diol (3.52 kg, 2.46 mol) and 200 g of THF. The solution temperature was lowered to 0°C. Fuming sulfuric acid (854.7 g, 9.11 mol) was then added to the flask. At this stage, the free acid form of [α,ω-(dihyrogen disulfato)poly(3-methyl-3-[(2,2,2-trifluoroethoxy)methyl]oxetane) is formed. The reaction was followed by end-group analysis, performed by ^{1}H NMR spectroscopy and ammonium hydroxide titration to a bromothymol blue endpoint. Once conversion exceeded 90 percent, the acid end-groups and excess acid were neutralized by addition to 25.2 weight percent aqueous ammonium hydroxide (708.2 g, 5.09 mol). The solution pH was monitored either by pH paper or meter to a pH of 7 to 8. The solution was allowed to stir at 0°C for two hours to allow for product formation. Excess salts were removed by vacuum filtration. The solvent was removed by rotary evaporation to yield a clear, viscous oil. The synthesis of 4b was conducted identically using 2b. Yields were typically greater 95 percent for 4a and 4b. The degree of polymerization was checked by ^{1}H NMR spectroscopy and found to be unchanged from the respective macrodiol.

Polymer Derivative Scale-Up

Scale-up of polymer derivatives requires careful consideration of physical properties and heat of reaction. The laboratory equipment has limitations. It can be a challenge to experiment with alternative process methods in the laboratory. On a large scale conducting the reaction at 0°C increases the process energy requirements, and sometimes such equipment is not readily available unless an investment is made. It can be expensive. In addition, low reaction-mass temperature can increase the viscosity, thereby reducing the heat transfer—i.e., increasing the batch cycle time. In such situations past experiences and knowledge base can facilitate scale-up.

Such reactions can be conducted at a higher temperature if the process can be modified and the exotherm controlled by adding an external heat exchanger. A higher temperature necessitates excellent mixing and process temperature control.

Since the oleum reaction is highly exothermic, direct addition to the reactor mass can create a localized high temperature. Even with good mixing and a chilled reaction mass, highly localized heat can discolor the product, and it may not be suitable for the intended application.

One of the ways to prevent the localized heat would be to feed oleum at the inlet of a pump that is associated with the reactor and used for liquid recirculation and/or emptying. A heat exchanger would be located at the outlet of the pump and would control the reaction exotherm to the desired temperature. Using this method, one will be capitalizing on vigorous mixing and will be able to control the exotherm. This would also improve the reaction rate. A properly designed pump and heat exchanger will also reduce the batch cycle time.

Since oleum feed and the resulting exotherm can be adequately controlled, the suggested stoichiometry can be fine-tuned. Excess oleum can be reduced, thereby reducing the ammonium-hydroxide usage as well as product washing time to remove the salts.

Commercialization of many PolyFox™ products using the guidelines outlined has been extremely successful. It has been possible to produce multiple products exceeding current and projected demand with minimal investment in the pilot plant. Initial estimates based on the laboratory procedures suggested that the company would require significant investment in a new plant to meet market needs. The pilot plant provided sufficient information, which could then be used to further improve the process and design a new plant requiring much lower investment compared to the design based on the laboratory synthesis methods. Depending on market needs, it is quite possible to have a continuous process for monomers. They can be operated in campaigns to produce the necessary intermediates.

BLENDING SCALE-UP

Blending of chemicals produces products that are primarily used as coatings for the decoration of a multitude of substrates. Development of these products starts in the laboratory. Once the product meets the desired performance specification, it is scaled up in a larger tank/pot/vessel (based on what the respective industries call a mix tank).

Blending of all liquid products is relatively easy, as the liquids are weighed and/or metered in the mix tank. Some of the components have certain mixing nuances, which have to be incorporated in the process. Unless there is miscalculation or gross error in weight, liquid blending, if done as developed in the laboratory, will produce the desired product. Imparting color to the blend of liquids to achieve a desired shade is a well-known practice.

Scale-up of the coatings in cases where solids are part of the formulation does require attention. Mixing and/or agitation are important and have to be completely understood. Solids must be completely wetted, mixed, and reduced in particle size to deliver the expected performance. Wetting of solids is a critical part of the mixing and grinding operation. If any thixotropic agents are part of the formulation, attention is needed. Their incomplete development can lead to performance problems. Colorant addition and color matching are an important part of the formulation.

Conservation of raw materials and strategic solvent use increase total process productivity, improve yield, and reduce the cost of goods—i.e., improve profitability. Exploitation of physical properties can influence and significantly improve the manufacturing technology. All these need to be capitalized.

SCALE-UP INFORMATION BENEFITS

Scale-up is an ideal time to select unit operations that will facilitate processing. The information gathered here can be used to select construction and gasket materials. Based on the process conditions, equipment design criteria are defined. Safety and environmental conservation can and should be a significant consideration in the overall process. Proper equipment selection and cross-fertilization of technologies can improve processing and product quality and reduce manufacturing costs.

REFERENCES

1. Bisio, A. and Kabel, R. L. *Scaleup of Chemical Processes.* John Wiley & Sons. 1985.
2. Leng, Ronal B. "Scale Up Specialty Chemical Processes Directly." *Chemical Engineering Progress.* November 2004;37–44.
3. Nadkarni, D. V. and Hallissey J. F. "Improved Process for the Preparation of 6-chlor-5- (2-chloroethyl) oxindole." *Organic Process Research & Development.* 2008;12;1142–1145.
4. Gurjar, M. K., Murugaiah, A. M. S., Reddy, D. S., and Chorghade, M. S. "A New Route to Prepare 6-chloro-5- (2-chloroethyl) oxindole." *Organic Process Research & Development.* 2003;7;309–312.
5. Nadkarni, D. V. "Improvements in the Preparation of Key Intermediates in Manufacturing Processes of Torcetrapib and Ziprasidone API," Organic Process Research & Development Conference. April 27–30, 2009.
6. Private communication with Dr. D. V. Nadkarni. June 12, 2009.
7. U.S. Patent 4,831,031. Pfizer Inc. May 16, 1989.
8. U.S. Patent 5,206,366. Pfizer Inc. April 27, 1993.
9. U.S. Patent 5,650,483. Aerojet-General Corporation. July 22, 1997.
10. U.S. Patent 5,654,450. Aerojet-General Corporation. August 5,1997.
11. U.S. Patent 5,807,977. Aerojet-General Corporation. September 15, 1998.

12. Kausch, C. M., Leising, J. E., Medsker, R. E., Russell V. M., and Thomas, R. R. "Synthesis, Characterization, and Unusual Surface Activity of a Series of Novel Architecture, Water-Dispersible Poly." *Langmuir.* 18;15;5933–5938.
13. Prepared by Dr. Richard R. Thomas, PolyFox™ Technical/Marketing Manager, OMNOVA Solutions Inc. Personal communication. June 25, 2009.

Ideas and Observations from the Author

In the ever changing business world, innovation is essential. This is very apropos in the chemical world as well. Pharmaceuticals are an integral part of the chemical world. As the economies of the developing world grow, pharmaceuticals could grow at about a 10% rate. With such growth, it is expected that the pharmaceutical market revenue could reach about $2.0 trillion in the next 10 years or sooner. Since pharmaceutical manufacturing technologies have laggard compared to the technologies practiced in fine and specialty chemicals, it is necessary to cajole pharmaceutical technocrats to innovate. I have expressed my ideas and observations based on the changing field.

PROCESS CENTRICITY IS THE KEY TO QUALITY BY DESIGN
Tuesday, April 6, 2010

It would be worth reviewing the future of pharma's most discussed acronyms (APT, QBA and QBD). The following are my interpretations of these acronyms. PAT (process analytical technologies) means various analytical methods that can be used to convey the state of the sample as soon it is tested. QBA (quality by analysis) is a methodology where the intermediates are tested by "off-line" sampling. The results tell us an "after the fact" state of the manufacturing process. Repeated testing is used to tweak the process until the desired quality product is produced. Such testing is the manifestation of a lack of complete understanding of the chemistry, process, equipment, and any and all variables that interact to produce a product.

QBD (quality by design) tells us that the people who have developed and designed the process have complete understanding of the process, equipment, and their interaction as the sample tested would meet specifications any and all the time. No intermediate sampling is necessary.

Around 2001, the above acronyms were coined for the pharmaceutical world. They were an instant buzz and synonymous with the current and future state of manufacturing. However, based on reading much of the published literature, I get the impression that "PAT" is considered a cure all and by waving this magic wand, we will produce a quality product all the time. However, this is far from reality.

Published literature also suggests many differing interpretations of these TLAs (three letter acronyms) and that could be the one of the reasons for very little progress toward QBD adoption.

Any analytical equipment that costs more than $30,000 (just a number) and requires an analytical chemist to operate and interpret the results for a commercial operation is expensive. As stated earlier, an intermediate process sample tells the state of the sample tested, i.e., is the process on track or not. This testing will not and cannot fix the manufacturing process automatically unless the analytical equipment delivers real time results and has feedback loops to control the process stoichiometry and operating conditions. In order to have this level of process sophistication one must have complete understanding of the chemistry, process equipment, and operating conditions, i.e., one has to move from "chemistry centricity" to "process centricity."

What is process centricity? Process centricity to me means moving away from "chemistry centric" laboratory practices and commercializing unit processes by applying appropriate unit operations. This would allow operating personnel to have complete command of the chemistry, process equipment, and operating conditions. They can create a process error, observe the process change, and correct the error in minimal time without producing off-spec product. Chemists and chemical engineers must incorporate this level of knowledge in the process. This would be perfection (almost) and would not require complex analytical methods to check the process and the product, as we would produce quality. We would achieve QBD, i.e., nirvana.

Since the majority of APIs are fine/specialty chemicals, their manufacturing practices are very similar. Due to dosage needs, the API annual manufacturing production volumes are significantly different from fine/specialty chemical volumes. Thus, it is necessary that we evaluate the current manufacturing practices. Process centricity might necessitate that we change/alter API manufacturing practices. We have to implement methods that are simple and based on good chemical engineering principles and practices. Technologies and methods exist to achieve this change, but due to "chemistry centricity" we have not made any significant progress.

As I have explained in a recent article and blogs companies most suited for implementing "quality by design" are the API producers and the formulating companies. They have to move away from "chemistry centricity" to "manufacturing centricity" during the technology transfer and incorporate methods that do not require any intermediate product sampling and analysis. This might not look or sound easy, but is the only way to produce a product based on "quality by design."

It is a bit disheartening for many that after almost 10 years we are still discussing PAT, QBA, and QBD. If more than 51 percent of the API producers and formulators stop intermediate sampling, QBD would become a way of future life in the pharmaceutical world. If this does not happen soon (let us say in the next two to three years) my conjecture is that PAT, QBA, and QBD will disappear from the pharmaceutical vocabulary like any other fad. That would be sad because we collectively would have failed to implement a good idea that not only will improve profits, but also might facilitate regulatory requirements.

ALTERNATE INTERPRETATION OF PHARMACEUTICAL TLAS (THREE-LETTER ACRONYMS)

Tuesday, March 16, 2010

Around 2001, Dr. Ajaz Hussain and his colleagues at USFDA (based on their experiences) suggested and encouraged the global pharmaceutical industry to improve their process technologies so manufacturing practices could be simplified and quality products could be produced with minimum interference. It has been a noble effort and led to the coining of two acronyms—QBA and QBD. I have seen considerable analysis and discussion on how to move from "A" to "D." I would like to restate these acronyms differently in a lighter perspective with the hope that we will improve our manufacturing technologies.

In the current state of manufacture of active pharmaceutical ingredients (APIs) and formulated products, we repeatedly check intermediates and the final product for quality. This practice has been appropriately called "quality by analysis" (QBA). It not only prolongs the manufacturing cycles and processes, but also impacts the whole business process through increased inventories of raw materials, in-process materials, and finished goods. The whole business process becomes complex and lowers profitability.

I would like to rename the acronym QBA and call it "quality by aggravation." Why aggravation? Meticulous sampling and analysis is necessary to ensure we follow protocol at each step. This prolongs and extends the batch cycle times. Everyone gets aggravated if agreed protocol procedures are not followed. An intermediate product not meeting specifications is either reworked or disposed of. All this costs money. In addition, it increases work in process inventories that affects cash flow. Additional investment could be needed to quarantine intermediates, and all of this lower profits. In doing all these we aggravate our life and it has often become our way of life.

Aggravation (mental and financial) in pharmaceutical manufacturing processes can be alleviated if we have complete command of the process chemistry and its unit operations so that we can produce quality product with minimum or no intermediate samplings. Since we will produce products based on "quality by design" (QBD), I would also like to rename this acronym to "quality by desire." We will have quality production. This will simplify the manufacturing process and life. Our desire to have an excellent process could significantly reduce our aggravation. It even could be eliminated.

Pharmaceutical products have to meet the strictest of the quality standards no matter how the products are produced. The question that needs to be put on the table is what should be our method of choosing to produce a quality product. "Quality by aggravation" or "quality by desire" are the two choices and we must choose one.

Most of us know the correct answer. All of us have a choice and can do what we desire. Since maximizing profitability is the ultimate goal, I am not sure why we have avoided the right path for so long. Is the industry waiting for disruptive innovation?

QBA and QBD methods are applicable to everything we do in life. When we desire anything, we make sure that our methods to achieve the goals are swift and simple. We always pick QBD. Our choice in the manufacture of pharmaceuticals and their components is "quality by aggravation" or "quality by desire." You choose.

HIV DRUG AVAILABILITY AND POTENTIAL MANUFACTURING OPPORTUNITY
Tuesday, February 2, 2010

The global spread of HIV/AIDS has been and continues to be a cause for alarm. Approximately 33.4 million people are estimated to be infected with HIV/AIDS and about 30% of this population can use the antiretroviral therapy (ART). Of this number, about 42% are getting the treatment. This could be due to pricing and/or its availability or a combination of both. There are methods and means to lower the cost and increase availability. I have used AZT (azidothymidine/zidovudine), an ART component, as an example to illustrate the need for manufacturing technology innovation that can lower prices and increase availability.

In general, the majority of active pharmaceutical ingredients (API) are manufactured using batch processes. This is due more to tradition rather than the process chemistry and economics. This holds true for high and low volume actives. Some of the actives due to their chemistry and volume have become commodity chemicals and are produced by continuous processes. However, batch processing is still the preferred method of API manufacture in China and India irrespective of their volume.

Under the Clinton Foundation HIV/AIDS Initiative (CHAI), the prices of some HIV/AIDS drugs have been negotiated with the suppliers from India and China. An AZT (300 milligram) tablet is priced at 13.3 cents.

Doing a reverse calculation and using about 80% tablet formulation yield, one can calculate the potential bulk selling price of azidothymidine. Using two 300-milligram tablets per day, about 2.4 million pounds of azidothymidine would be needed for about 4 million patients. At $25.00 per kilo, the cost of the API content would be about 1.88 cents per tablet. If the formulation, excipient cost, and profit were considered to be five times the cost of API (an estimate), the cost of a finished tablet would be about 11.25 cents compared to 13.3 cents from CHAI. Thus, the assumption of $25.00 per kilo for bulk API is not un-reasonable. All of the AZT in the above consideration is produced using batch processes.

In the coming years under the new World Health Organization guidelines, if the number of people needing treatment grows to 14 million, there would be a supply problem. The capacity of the existing batch processes would have to be increased to about 8.5 million pounds of AZT per year. Similar steps would be needed to increase capacity of other members of the ART cocktail. Another alternative is to develop and commercialize continuous processes.

A continuous process would not only increase throughput, but also lower the manufacturing cost and consistently produce product of high quality. The combination of improved manufacturing technology and throughput can easily lower costs by 20–25%. Based on the reported chemistry, it should be feasible to develop a continuous process. We should never forget that most of the actives are fine chemicals first and drugs second. Acceptance of this fact might facilitate development of better manufacturing technologies. If better technology could reduce the price of AZT from 11.25 cents per tablet to 9 cents, that would be a significant improvement.

A continuous process would have much higher throughput than the batch processes facilitating availability of the needed drug. Plants can be ramped up and down to meet the market demand. Three plants using continuous processes and operating at about 400 pounds per hour (24/350 at 85% on-stream-time) could meet the global demand of AZT and give the operators a higher profit margin. The strategy discussed above could be extended for other components of the HIV/AIDS cocktail. It would be a win-win.

A RADICAL APPROACH TO FINE/SPECIALTY API MANUFACTURING
Wednesday, January 20, 2010

The average wholesale price (AWP) of blockbuster drugs (sales greater than $1 billion/year) and dosage determine the quantity of API needed. This calculation can easily be made based on information in the public domain. For a 200 milligram dose blockbuster, a decent process for a certain molecule at $10.00 AWP would require about 25,000 kilos of API. The API need would change with different AWP and dosage. A poor process and yield would mean that its pollution (not just carbon) footprint is much larger. At $50 per kilo, the API revenue would be about $1.25 million dollars, which is miniscule compared to the total drug revenue.

Due to low volumes, the global API producers resort to the easy, traditional method that we are taught through our textbooks—batch process. The nature of the batch process diminishes/prevents any implementation of "quality by design (QBD)" methods and we resort to "after the fact" analysis and fixes, which cost both money and time. This situation is the norm for pharmaceutical fine chemical manufacturing.

Due to the price differential between AWP, the factory cost of API, and the tablet/capsule, there is little financial incentive for the drug wholesaler to invest in manufacturing innovation. However, the API producer and the drug formulator have a major incentive to improve their profits—being the manufacturing technology innovation leader. A production paradigm shift on the part of producers and formulators is needed to achieve that goal.

Creative incorporation of physical properties and unit processes, as well as manipulation of unit operations and modular plants can facilitate QBD. This will serve to ensure continuous processes producing quality product the first time and all the time. Modular plants can produce almost any combination of fine or specialty chemicals. Since the API volumes are low, they can be campaigned allowing different products to be produced by companies with proficiency or expertise in specific chemistries and/or methods. Entities with knowledge of alternative manufacturing methods can easily produce some of the actives using continuous processes. A properly designed facility can produce about 55,000 pounds of product operating 24/7 at 100 pounds per hour in about four weeks. A batch process can take longer and would require greater investment.

Fine/specialty chemicals such as 3-(diaminomethylidene)-1,1-dimethylguanidine hydrochloride, 2-[1-(aminomethyl)cyclohexyl]acetic acid, (RS)-6-methoxy-2-((4-methoxy-3,5-dimethylpyridin-2-yl) methylsulfinyl)-1H-benzo[d]imidazole, various fluoroquinolones derivatives, 2-[di(phenyl) methylsulfinyl]acetamide, 1-[(2R,4S,5S)-4-azido-5-(hydroxymethyl)oxolan- 2-yl]-5-methyl-1,2,3,4-tetrahydropyrimidine-2,4-dione are a few examples of what can be produced by batch processes. However, continuous processes using modular unit operations can also produce these products. One must be creative and able to effectively incorporate the nuances of physical properties and reaction kinetics into the manufacturing processes. The above chemicals are examples of anti-diabetic, anti-bacterial, pump protein inhibitor, anti-viral compound, and other disease curing actives.

Traditionally, in the development of pharmaceutical fine/specialty chemicals we get enamored with incorporating regulatory practices and guidelines before we have an excellent process that will produce repeatable and consistent quality product without "in-process" analysis of intermediates. This is like trying to fit a square peg into a round hole. For manufacturing technology innovation, we have to step out of our comfort zone. The North American automobile industry, for example, got trapped in its comfort zone with very discomforting results. Chinese/Indian or any other companies could be the "creative destructionist" and change the global playing field.

Alternative manufacturing technologies and methods will force process efficiencies and lower the pollution footprint. API manufacturers and drug formulators must take the lead in utilizing these methods. Since such methods would be innovative, we could also see the reduction or stoppage of job migration to developing countries. With the right technologies, cGMP would be a given.

WHAT IS JUGAAD (NEW MANAGEMENT FAD FROM INDIA)?
Thursday, December 10, 2009

In recent years we all have been reading and hearing about what the Indians have been doing in the areas of IT, chemicals, pharmaceuticals and petrochemicals, etc. They have accomplished it all through desire, creativity, and Jugaad.

A recent issue of *Business Week* covers "India's Next Global Export: Innovation" called Jugaad. So what does this new management fad mean? In simplistic terms, Jugaad literally means an arrangement or a work around, which must be used due to a lack of resources.

This definition is very apropos. All of us have exercised Jugaad in our lives and may not have realized it. It could be developing a better process or a product. It could be stealing a base in baseball, kicking a soccer ball to score a goal, or the topspin of the tennis racket to win the match, i.e., achieving an objective using whatever it takes. All of us have the creativity to achieve our goals and objectives. We have what it takes.

Many of us may not have heard of Mr. Michael O'Leary. He is the chief executive of Ryanair Holdings Co., the Irish no-frills airline. He is putting his Jugaad to practice. His creativity was evident when he started a one-car taxi company to legally use Dublin's bus lanes and cut an hour from his daily commute. This is Jugaad. Jugaad is not an Indian thing. It is everywhere. All of us have it. We just need to do things in a manner that simplifies things and processes.

Can we apply Jugaad to the manufacture of chemicals, pharmaceuticals, polymers, resins, and other chemical-based products? Yes, we absolutely can. By learning the fundamentals of chemistry, physics, and mathematics we have the knowledge base. Along with these fundamentals, if we just apply our creativity and imagination we will have the simplest and most cost effective processes producing the highest quality products. Customers will come back time after time and will generate sufficient and significant profits.

AN INTERPRETATION OF U.S. FDA GUIDANCE FOR PHARMA MANUFACTURE
Thursday, October 8, 2009

About five years ago the FDA issued its "PAT Guidance for Industry" (September 2004). The FDA's intent by issuing the guidelines was to encourage the pharmaceutical industry to improve and innovate its manufacturing practices so that they produce quality product the first time and all the time. Manufacturing innovation will be the lifeline for pharmaceuticals as their business model changes.

Regulatory bodies want the companies to move away from achieving quality through the current practice of "after the fact repeated analysis."

Achieving product quality the first time and all the time is the goal. This is not a difficult expectation.

The guideline is suggesting to the industry what needs to be done and the methodology for improving the processes. However, innovation has to come from within the industry rather than thrust upon them. The guidance is very legalese and can be interpreted in many different ways. As written it eludes more to drug formulation than to active pharmaceutical ingredient (API) manufacture. However, the rules of the game for API and drug formulation are the same, as quality is the ultimate goal.

If we clear the forest and the legal jargon, the FDA is saying "understand how the chemicals react, interact, and behave with each other and have a process that if operated at the desired process conditions should deliver quality product the first time and all the time" (basic tenants of chemical engineering, chemistry curriculum and process development). Anything short will not deliver first time quality product.

In product and process development, we need to understand the chemicals, their interaction, establish specifications, and use different analytical technologies to ensure that the developed process will deliver the expected product. If one is expecting that process analytical technologies will fix a bad process, that would be a gross error and expectation. Analytical technologies tell us the result rather than the path to the result.

Good manufacturing practices and continuous improvement is a must for all manufacturing. PAT guidance suggests that the pharmaceutical industry should have discussion and approval from the FDA on their existing process improvement plans. This is adding costs. The benefits of process improvement can be quantified but since the re-approval costs are not known, my conjecture is that the industry would not opt for any process improvement for their existing products as the costs could exceed the benefits. Estimated savings due to process innovations for the pharmaceutical companies are in the $100 to $200 billion range. I hope this is good incentive to innovate.

Pharmaceutical companies, due to first to market pressures and following regulatory directives and guidelines, are not able to apply good chemistry and engineering principles to have an efficient process that produces quality product. The current state of pharmaceutical manufacture is manifestation of our methods. If we are expecting it to change, pharmaceutical companies must take the lead. Innovation can happen for the products that will become generic in the coming years and for the new molecules that will be commercialized.

The development of innovative processes has to start during the process conceptualization and development. Even then it would require an effort as old methods and thinking would have to be discarded, which is not easy. The regulatory bodies will have to be flexible and encourage innovation. PAT guidelines and other guidelines are encouraging innovation but have too many constraints. I strongly believe that innovation can reduce regulation.

FINE CHEMICALS: QUALITY MANUFACTURING AND TECHNOLOGY INNOVATION IN PHARMACEUTICALS
Friday, September 25, 2009

A recent survey "Pharmaceutical Process Control: Is the Great Divide Growing?" makes one think and ponder about the direction of manufacturing technologies and process development methods in the changing pharmaceutical business model.

I found some of the answers to be conflicting. It is expected that the problems of technology inefficiencies should go away. However, sometimes costs and processes get in the way. The survey suggests that PAT and QBD could be mutually exclusive, and this was surprising since they cannot be.

For a chemical process to produce quality product complete understanding and incorporation of the physical properties of chemicals, their reaction chemistry, and interaction is necessary. Understanding facilitates development of an excellent process. These are the fundamental elements of QBD and PAT.

The survey raises the following questions:

1. Do the survey answers give the direction of the company as a whole or only the thinking of the participating staff? Is the staff opinion in sync with what management wants?
2. What is management thinking with respect to manufacturing and process technologies?
3. Are the survey questions such that answering "yes" to one part of the survey could result in an automatic "no" for the other part of the survey, i.e., consistency or lack of it?

My focus is on having the best process development and manufacturing technologies so that we can have a process that is safe, environmentally sustainable, and produces quality product first time and all the time without repeated analysis.

If we understand the fundamental elements, our creativity and imagineering should result in "state of the art" processes that will produce a quality product. Proper process controls are derived from such knowledge.

Unless we understand the fundamental elements, after the fact improvement effort (lean, six sigma, etc.) would not result in an optimum process. Actually, such an effort can be expensive. Incomplete understanding results in a less than optimum process. It will be an expensive investment as is the case in pharmaceutical manufacturing.

Knowledge of elements will facilitate incorporation and adoption of state-of-the-art and new technologies. Microreactors are the new "to be discussed" technology after the pharma acronyms. They are being touted as the next best thing since sliced bread.

For the past 10 plus years, "microreactors" have been a laboratory curiosity. A microreactors is simplistically a reaction space that also acts as an efficient heat exchange device. If used properly it can lead to an "efficient, green, and sustainable" process. They are a modified/enhanced nano-version of plate and frame heat exchangers, which have been commercial for 40+ years. Such exchangers have been primarily used as heat exchangers rather than a combination reaction and heat transfer space. They perform extremely well in their dual role. These and similar technologies have to be understood and their value capitalized. Such reactors have a place in the pharmaceuticals (specialty chemicals) and fine chemical world.

The use of innovative technologies and improvement of manufacturing practices is only possible if we understand the fundamentals and apply principles of chemical engineering for an optimum process. The effort is not expensive, and once incorporated we would see very positive results.

A PHARMACEUTICAL CHALLENGE FOR TECHNOCRATS
Friday, August 21, 2009

Pharmaceuticals have their own unique technology and pricing positions compared with other chemical products. Can we introduce innovative development and manufacturing technologies for the pharmaceuticals sector? The answer in unequivocal "yes!" We just need to understand the roadblocks and overcome them.

Since we want to live forever, we are willing to pay the demanded price for a drug. Our willingness to pay for long life, along with the monopoly during the life of the patent, has been the primary driver for setting drug pricing. Drug prices are set at the highest level the market will bear. Once the patent expires, brand companies move on to invent new drugs.

The above two factors ensure the desired profit margins for the pharmaceutical companies. Any costs due to regulatory mandates are passed on to the consumer. Thus, the need for product, process development, and manufacturing technology innovation has been minimal. Inefficiencies are an accepted part of doing business. Generics have followed ethical (brand) companies in their modus operandi.

Regulatory bodies have cajoled pharmaceutical companies toward innovation by creating PAT, CMC, QBD, and other TLAs. However, these cannot be forced or mandated unless some other event takes place, which will have a financial return. (We are familiar with the phrase "you can lead the horse to water but you cannot make it drink.") There has to be a solution for this dilemma. Only an "economic incentive" will result in innovation.

Latent blame for the lack of innovation is placed on regulatory agencies. This is unjust. The repeatability of quality at the active pharma ingredients (API) and the final formulated drug stages is mandated—as it should be. However, the "path to quality" should not be mandated. Companies should be held responsible for "quality failure." The penalty for quality failure must be severe. Companies should have the freedom to choose the "path to quality" as it is the road to innovation and creativity.

Providing manufacturers with the freedom to choose their "path to quality" is the equivalent of "stopping the sampling of intermediates" for quality. This will force everyone to "drink the water." Companies will save significant money, which will be additional incentive for pharmaceutical development and manufacturing technology innovation.

Stopping intermediate sampling could be encouraged and even mandated. It will happen only if we understand "everything about the raw materials and intermediates but were afraid to ask." I am quite confident that based on the education and training that chemical engineers and chemists receive they can become the proponents of "stopping the sampling of intermediates." With their backing we will arrive at the destination where the regulators want us to go. Technology innovation is not difficult and for the technocrats it is a most exhilarating experience.

We need to keep API and drug formulation as separate processes and that will simplify innovation. In general, many articles discuss pharmaceutical process improvements. These do not include API manufacturing process improvements, but only refer to formulation process improvements. McKinsey & Company in a recent report suggests that pharmaceutical companies have an opportunity that exceeds about $65 billion through productivity improvements in the drug formulation area. Based on my review of the API segment, I believe that the opportunity in the API sector based on yield, technology improvements, and conservation far exceeds $65 billion.

The question is: "Are the chemists and chemical engineers ready and willing to take the challenge?" I know the answer and it is "Yes we can." If we do, many of the TLAs will become irrelevant.

CHEMICAL ENGINEERING: UNDERSTANDING THE CURRICULUM FOR QUALITY MANUFACTURING
Wednesday, August 5, 2009

Chemical engineers are taught during their training that they will commercialize and/or operate a process that will produce consistent quality product all the time (without re-work) using a safe, sustainable, and an economic process.

To achieve these objectives, we review topics that teach us the understanding of the physical properties of material (raw material, intermediate, by-product, and the product) involved in the process. This allows us to understand their interaction in a reactive and/or a blending process. Chem. E. uses this information to commercialize a robust process.

If we have mastered the properties and the interaction of chemicals involved, we should be able to define the operating conditions of a process having the highest yield with the above defined process characteristics. We are also taught various unit operations that we can use, modify, and/or manipulate to produce a quality product all the time. If we are not able to achieve the objective of producing quality product using a safe and sustainable process the first time and all the time, we have to improve our understanding so that we can create the correct process.

If I translate the Chem. E. training fundamentals to acronyms, we are taught to develop and commercialize a quality by design (QBD) process. This is our "hippocratic oath." Anything short of this objective suggests that we need to improve.

Regulatory bodies have introduced few other acronyms in the pharmaceutical manufacturing. Interpretations of these vary and introduce variability. My question is: Are we trying to have the best pharmaceutical manufacturing technology or are we trying to conform to the current fashion crowd?

My interpretation of QBA, CQA, CMC, DS, and PAT is as follows. If my understanding is not what the "gurus" expect it to be, then please help with the correct interpretation.

- CQA (critical quality attributes): We need to understand the physical properties of the materials (raw material, intermediate, by-product and the final product) and how they interact with each other.

- DS (design space): Definition of the process operating parameters that have been identified by the developers, which if followed will produce quality product all the time.

- CMC (chemistry, manufacturing, and controls): Reaction mechanism, kinetics, and process controls that are understood and followed will allow production of quality product.

- PAT (process analytical technologies): This acronym is the least understood. It is believed that by having PAT, all of the process ills will go away. That is far from the truth. Analytical instruments will let the manufacturing and quality people know that the process has erred. However, it will not correct the problem and provide a solution to the problem. Only people who are familiar with the characteristics of the materials and chemistry can correct the process. Analytical instruments are an indicator and not the corrector. There is a difference between process control technologies and process analytical technologies.

- QBA (quality by analysis): It suggests that we have a problem and we do not meet quality. We have to go back and fix the problem so that we can produce the desired quality.

To summarize, the above mentioned acronyms are the fundamentals of chemical engineering curriculum. If we understand pieces/parts of the curriculum, then we should have a QBD process. The question then arises, why it is so hard to implement the fundamentals of chemical engineering in the manufacture of a pharmaceutical (API or a blend of API and excipients) or did I miss something?

CLIMATE CHANGE AND ITS IMPACT ON INDUSTRIAL PRODUCTION
Tuesday, August 4, 2009

Climate change is a new challenge and a major discussion point between the developed and the developing countries. Some sort of emission limits will be implemented as we move forward. There is a lot of posturing and both sides are making point and counterpoint.

Developed countries did not have emission restrictions during their growth. With the current demand to curb emissions, some curbs will be negotiated. Developed and developing countries are afraid of the curtailment of their industrial machine. In order to retain their industrial complex, developed countries will exert pressure. However, the developing countries, especially India and China, are not going to readily agree to any curbs. During Secretary Clinton's recent trip, India rejected the U.S. proposal for carbon limits.

A recent article title "India and Climate Change" takes India as an example and excludes China, though both present similar challenges for the developed countries. The article states "If (the) developed nations are held responsible for emissions that they historically contributed, oblivious to their impact on climate change, why shouldn't (the) developing nations take responsibility for producing generations of people who will generate emissions into the future?" Is it an indirect admission that the developed countries are afraid of curbing their economic growth and are afraid of the growth of the developing countries? It seems to suggest that since the developed countries control their population, they can keep emitting at the current per capita rate. Is it also suggesting that the living standards of the developed countries should remain high while the living standards of the developing countries should not? If this is the latent intent, it is not going to sit well with the developing countries.

Are we saying "Maslow's hierarchy of needs" is only applicable to the developed countries? Developed countries have had the developing countries as their market, but now that they are challenging us on our turf, we are not willing to accept the challenge. The game has changed and we will have to play by new rules. Their development and negotiation is going to be a challenge.

Recently, the International Council of Chemical Associations engaged McKinsey & Co. to suggest steps the chemical industry needs to take to curb emissions and still innovate. This study excludes chemicals that improve the living standards (including pharmaceuticals) and assumes gross savings from such chemicals to be zero.

I have concerns about this exclusion as we are excluding an important segment (pharmaceuticals about $800 billion revenue out of $3 trillion per year) that has a large carbon imprint. Pharmaceuticals (API and formulated products) present an opportunity to reduce their imprint. There is an opportunity to improve their manufacturing inefficiencies (low yield) and reduce their solvent use, thereby achieving an offsetting positive impact. Technology improvement will also reduce healthcare costs. An effort is needed in earnest.

Development and global sharing of the low carbon emission technologies might be the answer. Another choice for companies in the developed countries is to move their factories to the developing countries. Thus, they would not have to implement tougher emission standards. This is not a viable option.

In the last 15–20 years countries have become dependent on each other. What was environmentally acceptable yesterday will not be acceptable tomorrow. Since the global warming will affect us all, we will have to compromise and live with the new rules, whatever they may be.

RECYCLING COATINGS: AN ENVIRONMENTAL AND BUSINESS OPPORTUNITY
Wednesday, July 8, 2009

With today's environmental concerns about how to reduce greenhouse gases (GHG)/carbon imprint, an opportunity exists in the coating business areas that can appease many. This is through the recycling of coatings.

Recycling of coatings is a possibility and a challenge. The challenge comes from the perspective of the formulators and the raw material suppliers. Raw materials deliver the desired coating performance. If the raw materials can be used interchangeably to deliver the required performance, we can have the makings of easier recycling and better manufacturing (batch/continuous) technologies. Certain scenarios exist.

Kelly Moore, a California-based coatings company, is producing recycled coatings and selling them under the "e-coat®" brand. Their coatings must contain a minimum of 50% post consumer waste. This suggests that they have made an effort and succeeded in recycling. Thus, there is a distinct possibility that other coating companies will recycle.

Over the last several years, different methods and applications of surplus coating have been considered with sporadic success. Sustained success is needed to reduce the environmental impact of the coatings.

If the government mandates coating recycling through EPA regulations, it would be called meddling in business. However, the government can assist by creating an incentive program for companies that recycle. VOC credits present the best opportunity and any company's effort in recycling should be rewarded.

A joint effort will be needed to establish such a VOC credit program. Companies should decide how they will develop and incorporate the recycled material in their products. Companies have the knowledge base and the creativity to develop coatings that can have a significant amount of recycled material as a part of their formulation. Strategic and interchangeable use of different raw materials is the key for recycling. This would be a win-win.

PHARMACEUTICALS: WHAT IS HOLDING BACK QUALITY BY DESIGN?

Friday, June 12, 2009

We come across many TLAs and their number is increasing. What is a TLA? It stands for "three letter acronym."

In the regulatory world, TLAs keep us on our toes. In the pharmaceutical world, two TLAs are in vogue—QBA and QBD. Everyone associated with the manufacture of pharmaceuticals is familiar with these acronyms. But just to reiterate, QBA is product "quality by analysis" and QBD is "quality by design." QBA is the current tradition of the pharmaceutical manufacturing processes whereas QBD presents what the technology should be in the future.

Ongoing discussion suggests that there is a significant hesitation to improve technology. One must ask the question, "Why it is so difficult to move from 'A' to 'D'?" and I am sure many have. There seems to be a monumental hurdle/roadblock for the pharmaceuticals to move from QBA to QBD. I do not think there are any hurdles. We are just up against tradition. Since the traditions are entrenched in pharmaceuticals, we have accepted the current manufacturing practices. They have not been challenged. We are also afraid of the "regulatory gods." The move from QBA to QBD is very simple and the roadblock is staring at us. However, it has not been obvious to us. I define the hurdle/roadblock for the move from "A" to "D" to be "the isolation of intermediates of the reaction or the formulation steps." The mantra for QBD is "stopping isolation of intermediates."

If we isolate a reaction product after every reaction step or a mix after every formulation step to test the quality and the conversion yield, we are acknowledging that we do not have a complete understanding or control of the process step and its mechanism. If we did have the understanding, we would not be isolating the reaction step and/or blend intermediate and testing them for their quality.

The specialty/fine chemical industry by and large has a complete understanding and control of the processes. It does not necessitate isolation of the intermediates, as the quality is designed in the products. If we can achieve the same level of proficiency for the pharmaceuticals, we would move from quality by "A" (analysis) to quality by "D" (design).

In the pharmaceutical industry the move from "A" to "D," will be a major accomplishment in simplifying the manufacturing technologies and processes. It will not only improve process efficiencies, but also reduce the carbon footprint of the fine, specialty chemicals and the pharmaceutical manufacturing processes. It will reduce the cycle time for many batch processes and could nudge quite a few products to be manufactured by continuous processes.

Jumping the "A" to "D" hurdle is simple and easy. We just have to set our heart and mind to it. If it happens, my conjecture is the even the "regulatory gods" will celebrate.

PROCESS OF CONTINUOUS IMPROVEMENT AND PHARMA-CEUTICALS
Monday, June 1, 2009

In every industry, "process of continuous improvement" is a religion since it improves profitability. A recent article, "Drug CEOs Switch Tactics on Reform," in *The Wall Street Journal* discusses new strategies being developed by the pharmaceutical companies. Pharmaceutical CEOs believe that drug costs do not contribute to high healthcare costs. The following points are mentioned in the article:

1. Prescription drugs account for "just about 10% of the overall (health care) costs."

2. Reforms shouldn't force doctors and patients to choose a drug based on cost if the more expensive treatment would have a better outcome.

3. Drug makers have been pushing through hefty price increases. Prices for many drugs were up more than 15% in the first quarter from a year earlier, according to data from Credit Suisse.

4. Drug industry executives are worried about Medicare's authority to negotiate the prices for drugs dispensed through its Part D benefit. That could limit the prices pharmaceutical companies can charge.

5. Pharmaceutical executives argue that such steps (negotiated drug prices) would hamper drug makers' ability to pay for costly research into new treatments, saying "it would knock our legs out."

If healthcare costs are to be reduced, it has to be a full court press on every element of the costs, and that includes drug costs. Drug costs cannot and should not be excluded even if they are a small part of the overall costs. The pharmaceutical companies should make any effort to lower drug prices as part of their continuous business improvement process. Point #5 suggests that the drug companies want to fund the development of new drugs through raising drug prices only. If an effort is made to improve their R&D methods and manufacturing technologies, which is definitely feasible and possible, the pharmaceutical companies will not only have more funds to develop new drugs, but will also have higher profits.

It is well known that current drug manufacturing technologies and methods are inefficient. Effort needs to be made to improve manufacturing technologies. Improvement in API and drug formulation yield, e.g., from 60% to 90% might not seem to be a major improvement, but every dollar saved adds up. These savings might be in billions of dollars or euros and will be more than sufficient to pay for new drug research and development.

We all need to work together to reduce healthcare costs rather than saying the problem lies some place else. Suggesting that the problem is elsewhere is an indirect acknowledgment by the pharmaceutical industry that we do not believe in the "process of continuous improvement," and thereby cannot reduce drug costs. With the effort being made by every government to reduce healthcare costs, I hope the pharmaceutical companies are not saying that we have no room for such improvements and "do not tread on me."

Based on the fundamentals taught in engineering schools, every student will say that the current manufacturing methods can be improved. The real question is why such effort has not been made and what is blocking the path of "continuous improvement." It is well known that if manufacturing methods are improved, they will improve profit margins to levels that are much higher than the current levels and some of the savings can be passed on to customers to make it a win-win.

The question is, "Can and/or should an effort be made to reduce drug costs?" The answer is yes, and if someone says it cannot be done the question is why not.

PHARMACEUTICALS IN THE WATER
Friday, April 24, 2009

Every so often we read about how pharmaceuticals are being discharged into global water systems. It's good that we're being told and reminded that this is a problem we've created for ourselves. Unless these pharmaceuticals are removed from water, they will accumulate to a level that will have ill effects on both our bodies and our ecosystems.

There are two distinct issues here, and they really should be separated. Every article I've read combines the two issues—this makes it more difficult to find a real solution to the problem. The two issues are:

1. Pharmaceuticals in the water due to humans discarding them. There are no laws to control these discharges.

2. Pharmaceuticals from the manufacturing plants leaking into water. Regulatory bodies have guidelines and laws to control BOD (biological oxygen demand), COD (chemical oxygen demand), and suspended and dissolved solids to certain levels. There is no incentive for companies that abide by the rules to cut toxic chemical levels any further.

We can analyze and talk about the toxicity of pharmaceuticals and their ill effects on humans and eco-systems, but if there are no laws to control them, little will be done.

Talk, unfortunately, is cheap. Yes, the manufacturing process efficiencies need to be improved, but if I can make my profit margin and meet the water discharge regulations, there's no reason for me to spend extra money to ensure water safety. There is simply no prospect of a return on such an investment. Conscience does matter to a certain extent, but the economics drive these decisions.

Unless we make a concerted effort to fix this problem, we are going to see another Patancheru. The ball is in our court.

NANO AND PARADIGM SHIFT
Friday, March 27, 2009

When commercialized, the Nano car was the joke of nightly talk shows and most news broadcasters. Earlier this week reporters wanted to touch and feel it, saying "Oh my God, no radio, no air-conditioning and it is a tin can." Maybe it is. It might be bottom of the rung but it illustrates what is feasible. We have a new point from which to go forward.

It is a HUGE paradigm shift, not only for the automobile industry, but also for every manufacturing industry. It gets 55 miles per gallon and gets you from point A to point B. Yes, it has its deficiencies but we still all talk about it over a drink.

Nano shows us that "element human hu" can do unique things. It can go to point "x," which is out there beyond our imagination, if we put our mind to it. Is it the new iPod of the manufacturing industry? Maybe.

At the turn of the twentieth century, the four-wheel gasoline buggy fascinated us. Did we ever think in the 1980s that we would have laptops that could launch a missile? Most of us would say no. We have driven film photography to a technology museum.

We are now looking at the next generation of adventure. Human creativity is beyond control and Nano is a rendition of the possibilities of manufacturing and technology innovation. It should be celebrated. Hats off to the human element.

Can we do anything? Yes we can!

PHARMACEUTICALS AND RETURN ON INVESTMENT (ROI)
Thursday, March 26, 2009

Every reader is an investor. Investors know there has to be a good return on their investment irrespective of the place of investment.

We have all been taught that different risks necessitate different ROI. For "low risk investments," a ROI of 10–24% is suggested, 24% being in pharmaceuticals. The ROI range for "average risk" is about 15–40%. Again, 40% for pharmaceuticals. ROI for high-risk investments should be 24–56%, with 56% for pharmaceuticals.[1]

In the past few weeks, three major pharmaceutical mergers have been announced. Total investment is about $156 billion. If the total investment is equally distributed between the three companies and each would like to have a "five year ROI," then (due to high risk) one should expect a "before tax" return of about $20 billion per year per deal. Another way to look at earning $20 billion/year is that the each company will have to have 10–20 blockbuster drugs on the market beginning in 2010. Based on each company's pipeline, I just do not see such a gusher. Unless the acquiring players know something we do not know, I believe these are risky investments considering that less than 5% of drugs become blockbusters and past acquisitions and their assimilation have not been stellar.

I would like the readers to opine on the recent pharmaceutical investments, share their thoughts and what they think are the short and long-term options for pharmaceutical companies.

Reference

1. Valle-Riestra, J. Frank. *Project Evaluation in the Chemical Process Industries*. McGraw Hill;1983;433.

GLOBAL FINE/SPECIALTY CHEMICAL INDUSTRY AND ITS CHALLENGES: CURRENT SITUATION
Thursday, March 5, 2009

The global chemical industry is going through multiple transformations and the current economic environment is not helping. It needs to address the following:

1. How to react to the current slow down.
2. Determine their long-term prospects.

Companies in Europe and the U.S. innovated and developed many unique molecules that have improved our quality of life and lifestyle.

Products include pharmaceuticals, polymers, additives, flavors and fragrances, fertilizers, and list goes on.

Some of the old giants have disappeared. Recent re-factoring of the European companies to rationalize their businesses has caused more turmoil than it has solved as the companies are still losing money. Some are trying to find themselves and some have given themselves new names after reorganization. Some of the new entities have not found equilibrium.

Lack of growth (i.e., growth equal to GDP growth is no growth) has been a challenge (some segments have had higher than GDP growth, but many are lower). On the other hand, growth better than planned has been exhilarating. Both have impacted profitability.

As the world grapples with the current slow down, more so in the developed countries than the developing countries, the future looks murky. To conserve profits, companies have selectively shuttered their plants. This might be prudent for the short-term but mothballing plants might not solve the long-term ills.

The impact of the expiration of pharmaceutical patents and lack of new drugs in the pipeline will reposition the global fine chemical industry. We will begin to see a sea change in the second half of 2010.

What is the recourse for the future? Current markets for the chemical products can be categorized as follows:

- Slow or no growth (growth equal to or less than GDP)
- Growth (growth greater than GDP).

In the current economic downturn, the human and social impact of shutting down and/or moving R&D and manufacturing from the slow and no growth countries to the growth countries can have significant negative consequences. However, such moves might be necessary for the multinational companies. In the slowing global economy, due to political sensitivity moving from developed countries and investing in growth markets is going to take longer than the normal time and effort. Lack of rapid decision-making might further complicate strategy development.

Until a few years ago growth in the under-developed countries was slow and these markets could be supplied from the developed countries. However, with much higher growth in these under-developed countries, it has become necessary for the multinationals to fulfill the market needs either by opening R&D and manufacturing sites or collaborating with local partners. This poses an interesting dilemma for the multinationals. Should they consolidate their plants and supply the needs of the developed countries, if possible, from the plants in the developing countries and shutter their operations is the developed countries? This option has its own challenges—how to explain such moves to shareholders, including employees, and how to blend in the local culture and nuances.

Multinationals face another challenge in the developing countries. It comes from the local enterprises that have served the local and global markets. These enterprises might not be technologically strong, but is a matter of time until they could become fierce competitors.

More than 50 percent of the global population lives outside the developed countries. In the next few years, growth is going to come from these markets. They might not require the technologies currently used in the developed countries. Technologies to suit the local market preferences and environment might have to be developed. A joint collaboration between the local companies and multinationals can be a fast track option. Going it alone could be an option also. However, it would require understanding of the local markets. In addition, multinational companies will have to invest in technologies and capacities that are economic and can meet the market needs from fewer plants. This could be a challenge, but is necessary for survival.

Manufacturing of commodity (slow or no growth) products will move to the lowest wage countries. India and China could benefit from such moves. Only offset to such moves is the development of better manufacturing technologies for commodity products, e.g., plastic additives, flame-retardants, corrosion inhibitors, and rubber chemicals, to name a few. They have to be such that they offset the lower labor cost advantage offered by low cost countries.

The newest technology (growth better than GDP) products will be developed in the labs in developed countries and could be manufactured anywhere to serve their respective needs.

THE WORLD IS CHANGING FASTER THAN WE CAN STRATEGIZE AND IMPLEMENT
Wednesday, February 25, 2009

A fine chemical version of Chernobyl? Patancheru, India: An opportunity for Quality by Design and Environmental Sustainability

A study by Dr. Joakim Larsson, et al.[1] in September 2007 has suddenly become an eye of a storm in India.[2, 3, 4, 5, 6, 7] There are denials of the scientific study as it exposes weak links. For the long term, these issues have to be addressed. If the problems are not corrected, the area could be equated to the Chernobyl of the fine/specialty chemicals and pharmaceutical industries. There is a solution out of this quandary.

The solution touches the heart of the manufacture of active pharmaceutical ingredients and their subsequent formulations. Using Professor

Larsson's study, I have presented potential scenarios for the Patancheru problem and solution.[8] The process yield can be improved. Effort is needed. Depending on total ciprofloxacin capacity, which is a quinolone, the companies can collectively reduce fall out by 30–60 kg/day. This might not look like a big number, but based on total daily production this is big. Based on toxicity this is significant. Similar savings can be achieved on other quinolones and other drugs.

In the recent brouhaha ciprofloxacin has been identified as the culprit. Actually the problem is much bigger. There are other quinolones and active pharma ingredients being produced and formulated by many companies in Hyderabad and the vicinity. Not only are there producers of these products, there are suppliers of the necessary raw materials for these products in the area. Effluents from these chemical plants also discharge in water bodies in the area. Even if the effluent meets the established local standards of chemical discharge, no one has established the toxicity of every chemical that is trickling in the ecosystem.

If we want to salvage the Patancheru eco-system, we should establish toxicity levels of associated chemicals and use them rather than the current chemical limits to control effluent. We have to recognize that every active pharmaceutical ingredient is toxic to varying degrees and their toxicity kills the disease causing bacteria. The lessons learned from Patancheru could be applied globally.

WHY WE HAVE A PROBLEM

High levels of chemicals entering the effluent treatment plant point to inefficient manufacturing technology and low yields of the manufacturing process. Questions should be asked about why we have a problem and if the yields could be improved to reduce the effluent load, why it has not been improved. There is a simple answer to these questions and it encompasses the following.

1. Since high profit margins are made with the existing processes, there is no incentive to improve them. If the companies can meet the local water, solid, and air effluent standards, there is no need to worry about the eco-toxic or toxicity effect, as there are no standards.

2. The current processes with their current low yield produce a chemical that meets a certain impurity profile that has been approved the drug regulatory agencies. If the yield is improved, the producer should be able to reduce the chemical discharge load. This improvement could change the impurity profile of the active ingredient. Under the current regulatory laws of various countries,[9] the producer might have to re-qualify the higher drug produced by an

improved process for its performance and efficacy. This is an expensive and long drawn out process. In addition, processes might have to be re-audited. No one wants to invest any money in this effort.

3. Since the current processes are not efficient, the product quality is controlled at every intermediate step and this is called quality by analysis. The reason for low yield, i.e., high amount of chemicals in the effluent, is that the processes are not completely understood. Lack of complete process understanding and control can result is a product that is made on day 100 of one year and day 200 of the same year to be slightly different or might not meet specifications. If the product does not meet the defined specifications that have been filed and approved, the product could be reworked or disposed. These products and their intermediates are equally or more toxic and could leach out into the soil and water. Quality by analysis ensures high quality and this is expensive. These costs can be contained or eliminated if we have repeatable processes.

Problems identified by Professor Larsson do exist in many areas of the world where active pharmaceutical ingredients are produced. However, they have not been studied. I am sure we will find similar problems.

SOLUTION

The only solution out of the dilemma in Hyderabad is to improve the process manufacturing technologies. This has to be done for the short and the long haul. If the manufacturing processes can be improved, depending on the total capacity of the quinolone plants in the Patancheru area, significant quantities of ciprofloxacin can be recovered as a product instead of going to wastewater and solid disposal. It will make a big impact on the local ecosystem. Ciprofloxacin is one of the many quinolones being produced in Hyderabad. The plants producing ciprofloxacin also produce other quinolones. In addition, we have to recognize that there are ancillary plants in the area that produce raw materials for these products. Their effluent is part of the wastewater and solid sludge system. There are many other actives produced in the area and their levels have not been tested.

Manufacturing technology improvement is the only solution to reduce chemicals in the eco-system. Drug regulatory agencies have lately suggested that manufacturing improvements should be done. Formulators and producers of active pharmaceutical ingredients claim that there are hurdles of bureaucracy and insufficient ROI. I have difficulty believing that. Meeting chemical effluent standards at Patancheru would be the first step. Unless effluent toxicity standards are established, not much will change. Intervention is needed to rationalize toxicity and address the "Patancheru problem." We have to maximize our effort to improve manufacturing technologies.

References

1. Larsson, D. G. Joakim, de Pedro, Cecelia, and Paxeus, Nicklas. "Effluent from Drug Manufactures Contains Extremely High Levels of Pharmaceuticals." *Journal of Hazardous Materials*. Volume 148;Issue 3; 30 September 2007;751–755.
2. Mason, Margie. Associated Press, Drug Waste Creates Highest Disaster Zone in Andhra. *Times of India*. January 27, 2009.
3. Deshpande, Rajeev. "TNN, PMO Orders Testing of Patancheru Water." *Times of India*. January 28, 2009.
4. Kolanu, Manjula. "TNN, Officials Sleep as Pollution Sinks Patancheru Greens To Step Up Anti-Pollution Drive." *Times of India*. January 29, 2009.
5. "Independent Lab to Test Patancheru Water." *Times of India*. January 31, 2009.
6. "Silent Streams Turn Patancheru's Sorrow." *Times of India*. January 31, 2009.
7. Drug Traces in Patancheru Wells. *Times of India*. February 17, 2009.
8. Malhotra, Girish. "Pharmaceuticals, Their Manufacturing Methods, Ecotoxicology, and Human Life Relationship." *Pharmaceutical Processing*. November 2007;18–22.
9. Link to global regulatory bodies.

WHY HAVE THE FINE AND SPECIALTY CHEMICAL SECTORS BEEN MOVING FROM DEVELOPED COUNTRIES?
Monday, February 9, 2009

I am sure we all have been wondering about the shift. I have my own rationale, and potential solution I would like to share. I believe there are opportunities to innovate and we can capitalize on them.

I speculate and believe the industry has moved for the following reasons:

1. Environmental law
2. Health and safety laws
3. Significantly lower labor costs in third world countries
4. We did not invest in technologies to improve processes.

In the early 1970s, developed countries were adopting new environmental protection laws. They seemed unrealistic and unachievable (this is based on my being at a state EPA) to some. Many complied. For some it was easier to shut down rather than invest in the complying technologies. In my view, improving process technologies was a missed innovation opportunity. The chemicals were needed and since there was a need, companies in China and India filled in the supply gap.

India and China also had the advantage of rupee/yuan/dollar parity. This made investments in their country cheaper.

Laxity of health and safety laws persist in developing countries. The associated expenses are low compared to the developed countries. On recent trips, I saw workers with open toe shoes, without safety glasses, wearing street clothes, and eating meals on the operating floor. Human life needs to be valued as an asset.

Environmental laws are comparatively lax also compared to laws in developed countries. Thus, the respective investment in pollution abatement is lower. I have seen multi-colored water bodies next to the plants. Abatement of eco-toxicity is not a high priority. In the developed countries endocrine disruptors have been found in the drinking waters. I am sure these and other chemicals exist in the water in the developing countries also, but the scale is different.

Labor costs in China and India are magnitude levels lower than the costs in the developed countries, e.g., a plant operator in India could be paid $200 per month (we have to recognize high yuan/rupee/dollar parity) compared to $4000.00 per month or more in the U.S.

The combination of the above factors has resulted in China, India, and some of the Eastern European countries making fine and specialty chemicals to feed the insatiable need for these chemicals in the developed countries.

As time progressed, these suppliers found that their products were being used to produce the active pharmaceutical ingredients or other higher valued products, i.e., moving up the supply chain. These companies also moved up the supply chain. These have resulted in additional plant closures in the developed countries.

With time costs in the developing countries are increasing as they incorporate better safety, health, and environmental laws, but are not to the levels of developed countries. They still have a price advantage and customers willing to purchase their products.

Now we have a situation where many of the pharma APIs and other strategic drugs and products are coming from China and India. This is discomforting as expressed in a recent *New York Times* article, "Drug Making's Move Abroad Stirs Concerns."

We have to recognize that pharmaceutical and other companies are buying products (APIs, intermediates, and fine specialty chemicals) from the companies in India and China who are alleged not to be playing by the rules companies in the developed countries have to live by. It is a supply and demand question and rules in every country are different.

Can this be reduced, prevented, or stopped? Do we have a way out of this quandary? Yes we do, but it would require an effort. We have to have total involvement of the suppliers and buyers, which might not be easy. If such an attempt is made, I hope it would not turn out like "the Doha WTO negotiations" as many companies/countries have a lot to lose and/ or gain. I doubt if any trade organization can influence any country's environmental, health, safety, and pay scale policies. Those changes have to come from within. Maslow still rules.

SOCMA, CEFIC, and other organizations could identify the highest imported chemicals, APIs, or formulated products. Interested companies in the developed countries could develop technologies for these products that will offset the cost advantages of the imported products and convince the companies in the developed countries to buy their products. Every advantage perceived or otherwise from the developing countries will have to be offset by cost and quality through better technologies.

Partial protectionism under a "strategic defense initiative" could be a temporary alternate for certain chemicals or drugs. Such a program cannot be government subsidized. This could give interested companies a "time window" to develop better technologies. With many countries now part of WTO, such an initiative is not going to sit well with many countries, companies, and organizations.

Competing technologies that will offset the costs due to local wages, environmental, health, and safety rules and methods is the only answer. If this does not work, safeguards leading to a continuous supply of strategically vital products can be implemented, but there are costs associated with that strategy. A win-win strategy needs to be developed.

COMMODITIZATION OF DRUGS
Wednesday, February 4, 2009

Until 2006 low cost drugs could be purchased outside the U.S. only. Wal-Mart started to sell 30-day and 90-day supplies at $4.00 and $10.00, respectively, in 2006. Recently other pharma-sellers have joined to serve the growing market. These prices were unheard of before 2006. Even at these prices respective members of the supply chain (producers, formulators, and HMOs) are making "good" margins.

Commoditization had begun in 2006 and we did not realize it. With the current global economic downturn, an ever-increasing aging population, and economic upswing of under-developed countries demanding common ailment drugs, the commoditization pace has accelerated. As we go forward the number of drugs in the 30- and 90-day pool will increase. With the larger customer base, the annual volume for many of the active pharmaceutical ingredients (APIs) will increase.

Fine and specialty chemical companies (e.g., BASF and Albemarle among others are producers of ibuprofen and naproxen [non-steroidal anti-inflammatory drugs: NSAID]) and generic drug companies are producing common cure APIs (i.e., specialty fine chemicals that have disease curing value). Many generics formulate various dosages for sale. Most of the drugs U.S. pharma-sellers are offering for sale are being produced and formulated outside the U.S. Big pharmaceutical companies are not involved in these programs.

As more brand name drugs become generic and the volume of generics increases, entrepreneurs, existing and new, would want to take advantage of the business opportunity. Market economics and desire for profits will result in the development of better processes and movement from batch processes to continuous processes. Better process technologies will reduce the costs of active pharmaceutical ingredients (APIs), resulting in higher profits for the members of the supply chain. All this will result in higher profits and increased commoditization of the "off-patent" drugs.

It is expected that successes of better process technologies for the generic APIs might result in better manufacturing technologies for the ethical/brand drugs, also thereby increasing their respective profits. It is possible that lower costs from better processes for generics and ethical drugs might not necessarily be passed on to consumers.

"BAIL OUT OR HAND OUT" IS NOT THE ANSWER, BUT INNOVATION AND CONSERVATION IS
Thursday, January 1, 2009

Recent turmoil in the financial markets is taking its toll globally. However, this once in lifetime event is also giving us a message and presenting us with an opportunity. The message is "It is time to have innovative technologies that also conserve our resources."

The U.S. automobile industry lost its focus when it quit innovation in the 1960s and were not farsighted to raise the fuel efficiency. They fought tooth and nail against raising gas mileage standards. The Japanese came with better quality, pizzazz, and hybrids.

But Detroit thought it was not a good idea to have a "better idea." Lately we have read about plants of many chemical companies being shuttered for lack of demand. We will probably hear more such closures before things come back. Bankruptcies would be there also.

I wonder if the closures are a reflection of not having the best technologies to manufacture the products. Had the technologies been such that the feed rates could be lowered or increased to meet the prevailing demand, plant shutdowns could have been avoided. A lack of best methods suggests there is an opportunity for better manufacturing technologies. Better equals high conservation, i.e., produce more from less.

Pharmaceuticals, which are disease-curing chemicals, cannot think conservation when they are able to make their profit margins on "human desire" to extend their life. Poor yields, high in-process inventory, and quality checks only on every milligram suggests significant opportunities. Consumers pay for every inefficiency, e.g., inventories, poor quality and costs related to inefficient use of their raw materials. In 2007/2008 we saw loss of employment and knowledge base accelerate. When a pharmaceutical company can close more than 50 plants, it suggests that companies have technologies that need a total overhaul. They need to develop and implement technologies for their survival.

Their blockbuster model is dying on the vine and their new product pipeline is heading from gusher to a trickle in the next few years. Pharma needs to create a new business model. Due to the toxicity of their chemicals, pharma needs to improve their manufacturing technologies to levels better than "non-disease-curing" chemicals. Higher yields mean higher profitability and less effluent or/and emissions in our ecosystem.

On my recent trip to China, I saw electrically charged bikes moving around town. Similarly, in Europe and China they have low cost and simple solar water heaters on their rooftops to provide them with hot water for their daily use. Rooftop heaters do not look aesthetically bad, but they tell us the inherent character of the inhabitants and their nature to conserve and use nature's gift of the sun's heat. A missing rooftop heater suggests a "missing link." Communities in the U.S. have prevented such installations with the thinking that they look ugly and will lower real estate value. Aesthetics seems to be more important than conservation.

Chinese company BYD is introducing an electric car and Indian company TATA is introducing $2500.00 car. This suggests that innovation is possible if we step up to the challenge.

We need to move from "consumption zealots" to "conservation zealots." Conservation and preservation will not result in any hardship but will lead to innovation that will improve profitability. The present slow down is the best time to innovate and we need to spend effort so that we can reap benefits in future.

IS "CREATIVE DESTRUCTION" THE WAY TO GO FOR THE PHARMACEUTICALS?
Friday, December 12, 2008

The world is seeing the automobile (premier) industry requesting salvation from the government after they drove themselves into a ditch. Are the pharmaceuticals heading in the same direction?

A recent *Wall Street Journal* article, "How Detroit Drove Into a Ditch" is an excellent review of the auto industry. It clearly suggests that they lagged innovation and are suffering. As admitted recently by the management of General Motors, the only way out is to innovate and to do it in a hurry if they want to survive. Only time will tell, but based on their past record, the future looks bleak.

Pharmaceuticals have lived on the "blockbuster" model and have won. One player on the team has led them to victory for many years. Now it is time for the whole team to play together.

Unless R&D and manufacturing become strong, ethical pharmaceuticals cannot compete in the global market. Marginally better drugs and personal medicines will not generate the revenue stream once the patents have expired. It is time to compete on a global scale, i.e., serve the needs of 6.2 billion by serving across the globe rather than a small percentage of the population. Manufacturing and R&D need to innovate.

Dr. Severin Schwan, chief executive Roche Holdings AG, is correct. It is time to change the business model. Are pharmaceuticals the antithesis of creative destruction? I do not think so. We need to innovate for long term survival.

IS AUTO BAILOUT A PRELUDE FOR OTHERS TO ASK FOR HELP AND AN ADMISSION OF A "LACK OF VISION?"
Sunday, November 23, 2008

Headline: "An Auto Bailout Would Be Terrible for Free Trade." Does anyone really expect other countries to ignore our subsidies? The American automobile industry gave the world automobiles and held everyone in awe. However, the dire straights of the automobile industry suggest that it never thought much about the future. German cars were always considered a luxury and quality product and have never been considered a threat for the mass producers. It was the Japanese followed by the Korean carmakers that really changed the playing field by bringing quality from the get go. Their quality, styling, and innovation were the first threat to the survival of the U.S. automobile industry. However, the U.S. automobile industry has been slow to catch up. A recent article in *Wall Street Journal*, "How Detroit Drove Into a Ditch" gave an excellent overview of how the industry has arrived at its current state.

Japanese brought quality to the masses of the world and the world jumped on quality without paying luxury prices. The world was hungry for quality and fuel efficiency and did not get it from American carmakers.

Now the American carmakers are in trouble and asking for government help. If they are given a straw would they ask for more? Would other industries that are not able to compete with quality and cost ask for government help and protection? It is very possible.

Why are we here? Blame lies squarely on the management of the companies for their lack of foresight in innovation and delivering fuel efficiency to customers. Whatever happens in the auto world, it will work out for the good of the country. Better management is the answer. Asking for a government handout is taking advantage of the current economic woes rather than being responsible for their own inabilities.

If we step back and see similarities with any other industry that is experiencing some turbulent waters, it is the ethical pharmaceuticals.

Since 2005, the ethical pharmaceuticals have been facing head winds with their age-old "blockbuster" model. They will lose about $60 billion in revenue in the next three to four years. They have very little in their pipeline. They are scrambling to determine how they can sustain their revenue growth. The generic pharmaceutical companies are also challenging them on their turf.

Will pharmaceuticals be next in the handout line if they cannot solve their challenges, i.e., start growing their revenue with new drugs? History is repeating for the pharmaceuticals as it did for the chemicals, textiles, and steel industries. Chemical and textile industries have mostly moved overseas. Steel industry innovated its technologies to survive. Maybe the time has come for the pharmaceuticals to innovate their R&D and manufacturing technologies, which they have acknowledged need attention.

IS PHARMACEUTICAL CONSOLIDATION ON THE HORIZON?
Tuesday, November 4, 2008

Recently I had opined the following. Since then I keep reading views at other websites *(http://www.pharmatimes.com/WorldNews/article.aspx?id=*

14672). I still believe that a consolidation is needed to quickly fill the pipeline. Industry is vulnerable and the only reason venture capitalists have not moved in is due to the current credit crunch. Once the monetary crunch eases, we should see the beginning of consolidation. Increasing layoffs are suggesting that financial preservation is a must but it is coming at the expense of the basic knowledge base which is going to be difficult to replace. R&D and manufacturing technologies need to be brought to twenty-first century. Even with that, the basic business model will have to be revised. With increasing global effluence, market size will increase. In the increased market size the need for generics will be higher than the ethical drugs.

If one sifts through and compiles the news about the pharmaceutical companies, a clear trend with respect to their shifting business model starts to emerge. Slowly but surely, major pharmaceutical companies are behind the scene inching toward being a combination of "Blockbuster, bio-tech, and generic" models. This is their last and only hope. Merck is experimenting with a new business model of selling a patented drug (Januvia) at one-fifth the U.S. price level in India. Glaxo is venturing into South Africa and Egypt. These are undeclared secrets. Daiichi Sanyo has bought Ranbaxy. I am sure others are in the works.

I believe ethical pharmaceuticals are not very clear about what they want to be. As a result, they are dabbling with every opportunity they see, i.e., riding many boats with the hope that one will take them to the promise land. If they clearly define their mission, they might just need one big and strong boat (it could be a combination of blockbuster, bio-tech and generic) to take them to the goal. This will allow them to properly focus their attention.

Competing with the generic producers is going to be a challenge for the ethical producers. Their knowledge base is shrinking through layoffs. Their manufacturing technology is not current. If the ethical companies want to serve ethical and generic markets, they will need very efficient manufacturing technologies that can offset the generic producers' cost advantages. They can achieve this by collaborating and/or acquiring Indian or Chinese companies. With the globe shrinking, the second option is more likely. If this happens, it will lead to an eventual global consolidation in the pharmaceuticals."

RESHUFFLING OF THE GLOBAL PHARMACEUTICAL DRUG DECK
Tuesday, July 22, 2008

In the last five weeks we have seen the global generic pharmaceutical playing field change with the acquisition of Ranbaxy and Barr.

Until recently, the blockbuster model has worked for the ethical pharmaceutical companies, but with about $80 billion of ethical drug patents expiring in the next four years and with not much in the pipeline of major pharmaceuticals, their business model needs a re-look.

Major pharmaceuticals need to consider a strategy that would allow them to develop new drugs and also serve a large market that needs low cost drugs, whether they are patented or otherwise. It could be the last opportunity for the majors to get in on the generic business (similar to Novartis).

Recently Merck U.S. took a bold step in this direction without declaring a shift in its blockbuster model (Merck's low-priced diabetes drug might change a few rules). If they are successful, it would have other majors consider similar options. Acquisition and assimilation of the generics could be an option also.

In order for Merck to have success in selling their patented drugs at a lower price and maintain their profit margins, they will need to lower their API manufacturing and formulation costs. This will require a complete overhaul of their manufacturing technologies, i.e., moving from "quality by analysis" to "quality by design." Depending on Merck's success, other companies could follow the lead. This would be a giant leap for the pharmaceutical companies from their current manufacturing practices.

DRUG SAFETY, SIDE EFFECTS, THE FDA, AND ITS CHALLENGES
Sunday, March 9, 2008

Recently, the USFDA announced the "Safety First" program for the drugs that are on the market. WSJ reports, "Top Food and Drug Administration officials said this week consumers should expect to see more advisories and warnings from the agency about drug side effects."

It is an interesting and an intriguing program. I am not sure of its efficacy and how it will help consumers. The side effects of drugs that are commercial are public information and available. If one is expecting that the database is going to list all side effects, it is not going to do it and is going to come short of what everyone will expect.

We need to revisit and understand what drugs are. Drugs are toxic specialty/fine chemicals. Fine/specialty chemicals, to a chemist, are organic molecules that are mostly heterocyclic ring/s with nitrogen, sulfur, halogen, phosphorous, and/or oxygen incorporated in the ring/s and/or in their side chains. It is very likely that they have unsaturated bonds.

Drug evolution, development, and the regulatory review process leads to the introduction of many drugs. I am sure during the development process close attention is paid to how the drug will interact with the human body. There are checks and balances in place and only the drugs that have no or minimum ill effects enter the approval progression process. I do not believe that the interaction with every possible drug that humans take can be identified and quantified.

Developers and/or the regulators do not know how an unsaturated complex organic molecule is going to interact with another unsaturated complex molecule/s and acid/alkali of the human body. I do not know if anyone can speculate and/or conjecture how the molecules will break down and possibly recombine to create a new complex molecule in the human body. The only way to make a scientific conclusion is to actually study the effect of a combination of drugs.

Since the interaction of drugs is happening in the human body, the resulting chemicals cannot be sampled and studied for their good and/or bad effects. We all know that the human body is a well-controlled reaction system. Every bad effect on the human body is manifested by an illness, which is called a side effect.

Why did the USFDA take on this task? They are the "food and drug safety patrol" and this additional task is being taken on to placate its critics. Everyone eventually is going to treat the FDA database as gospel and an indicator of all ill effects. It is going to come up short on expectations and the FDA again is going to be blamed.

In addition, I do not know how they will carry out this enormous task when there is not sufficient money and/or manpower to handle higher priority tasks. This task is impossible at best. It is a lose-lose situation at best.

CHALLENGES TO ETHICAL AND GENERIC DRUG PRODUCERS
Sunday, December 23, 2007

If the current scenario of "drying of the blockbuster pipeline" and generics increasing their market share holds, we could see most of the API manufacture, formulation, and clinical testing moving to low cost countries. Since the "laws of economics" prevail, this could be considered inevitable. Ethical and Generic companies have to develop and implement strategies that could give them a competitive edge and allow them to move forward on their chosen path. Since ethical and generic producers are adversaries, it would be interesting to see the playing out of respective strategies. It is a real chess game.

ETHICAL PHARMACEUTICALS

Major pharmaceuticals have developed and commercialized blockbuster drugs. However, they have not retained these drugs in their portfolio after the patents expire, as they have been busy developing new drugs. Producing "patent expired" drugs has not been part of their strategy.

Due to high profit margins, generics have taken over the patent expired drugs and have lately made every effort to take over the patented drugs through litigation. With aggressive entry of generic producers from Israel, Iceland, and India, the turbulence in the pharma field has dramatically increased.

With the drying of the blockbuster pipeline, escalating clinical trial costs, and the relentless pressure of generics to capture the market, ethical drug producers are trying to implement strategies to reduce their costs and retain their stronghold on the drug development chain. Some of the strategies being implemented include the following:

- Outsource drug development
- Outsource active pharma ingredients (API) manufacture and formulations
- Synergize small molecules and/or biotech combinations
- Acquire small biotech developers
- Whatever else works, i.e., collaborations.

Some of these strategies might work as a short term fix to retain profits. However, the long-term impact of these strategies is going to be significant. The biggest consequence is going to be the shift, disappearance, and/or reduction of the knowledge base from "major pharma" companies to the outsourced companies. Since the "outsourced companies" are in low cost countries, they have the dual benefit of the above relationships. It makes them intellectually and financially stronger to become formidable generic competitors. We are beginning to see this happening.

Generic Pharmaceuticals

Generic pharmaceuticals are enjoying what I will call the best of all worlds. They are basking in unprecedented growth. I do not believe any of the financial analysts and pundits would have predicted this in the beginning of 2005.

Customers would like to have drugs at lower prices. Generics are able to fulfill this need in every market and as a result the demand for generic drugs has increased. This surge has increased generic business dramatically in recent years.

They have used these profits to grow organically and acquire sites that are being shed by API producers and formulators at significantly low costs. They have also been the beneficiary of technology and intellectual property that comes with these acquisitions.

Strategies[1] being implemented by the generic companies are unconventional and this is causing additional turmoil in the pharma field.

FUTURE AND STRATEGIES

Pharmaceutical companies have achieved handsome profit margins by inventing new drugs and by producing generics. Customers have paid for every inefficiency in the development, clinical testing, manufacturing, and supply chain. Since the pharma companies have been able to make respectable profits, they never saw a burning need to minimize the costs of each step. Everyone has been comfortable in their respective arenas. However, the drying of the blockbuster pipeline and Generics trying to encroach on the playing field of Ethical companies is changing the market dynamics.

The price we pay for drugs in the U.S. and some other countries is not market driven but driven by what the market can bear. Many consider these prices high and are getting low cost drugs any way they can, e.g., from Canada, Mexico, imports, and/or the Internet. This has led to considerable debate and discussions as healthcare costs increase. Wal-Mart and a few other companies are offering drugs at low prices. This puts pressure on the companies in the supply chain to make an effort to continuously lower their costs. Companies will have to consider and implement new strategies.

If the major pharma companies are not able to develop new blockbuster or biotech drugs, they could start making generic drugs. This could lead to consolidation and the formation of "mega" companies. My definition of a "mega" merger is a combination of an ethical and generic company to be players in both markets. These mega companies not only will develop new drugs, but they also will have to make every effort to retain the patent expired drugs as part of their portfolio. If this happens, every step of the supply chain (especially manufacturing technologies) would be critically evaluated and methods implemented to reduce costs. The business model of mega companies could be a combination of market and consumer driven companies trying to maximize their market share. This should reduce global healthcare costs.

The government of India[2] has announced an innovative drug discovery program combining global IT firms (Sun Microsystems), researchers (Royal Society of UK, Imperial College of London, Medicine Sans Frontiers, etc.), companies, and young minds at India's scientific laboratories to invent drugs at a fraction of the cost of a multinational company (MNC) developed drug. An open platform of drug research like Linux development is an interesting and innovative concept and path. Success here would genericize and commoditize pharmaceuticals and add additional pressures on pharma companies to implement technology improvements to reduce costs. Other business models will emerge and it should be interesting. I expect that better than 50% of the pharmaceutical market will be a commodity market in the next five years and we will see prices drop.

References

1. Generic Antidepressant May Affect Wyeth (*http://www.forbes.com/feeds/ap/2007/12/20/ap4461011.html?partner=alerts*) accessed December 20, 2007.
2. Government to Rope in Young Minds to Invent Cheaper Drugs (*http://economictimes.indiatimes.com/News/News_By_Industry/Health-care__Biotech/Govt_to_rope_in_young_minds_to_invent_cheaper_drugs/articleshow/2635842.cms*) accessed December 20, 2007.

THE YEAR 2007 IN PHARMACEUTICALS
Tuesday, December 11, 2007

The year 2007 might be remembered as an eventful year for the pharmaceutical industry. I have not kept a tally, but the number of layoffs and plant closures are significant. I do not believe anyone would have thought about this five years ago.

The pharma industry has had a myopic vision. They have believed that the pipeline will always be full and the "generics" will live on crumbs. They never invested in upgrading manufacturing technologies. Simpler technologies could have prevented generics from taking over the crumbs. Crumbs are now becoming stronger, bolder, and they will haunt the majors.

Unfortunately, the pipeline is sputtering and has leaks. Now the majors do not know what to do. The world needs cheaper medicines to live and generics can fill the need.

Jacob Schumpeter of Harvard said that industries go through "creative destruction" and it is a necessary part of the progress. Is the pharma industry heading that way?

THINGS WE KNOW ABOUT DRUG PRICES BUT ARE AFRAID TO ASK
Monday, November 19, 2007

Recently, an article presented a perspective titled, "Cost per Day/Cost per Kilogram: What's the Right Price for an API? That Depends Whether You

See Things Like a Pharma Company—or Like a CMO (contract manufacturing organization)."

This article illustrates why drug prices are high. Drugs are competitively priced with similar molecules. I do not believe anyone has asked the customer what she/he can pay or should pay. Customers want to live and they will pay the demanded price.

Due to the large price differences of medicines in the U.S. and many other countries, people explored alternate sources for their medicines. The Internet made that easier. This led to the import of drugs that could also be counterfeit and illicit. Hopefully, Wal-Mart's offer of $4 for a month's supply of generics will reduce these imports and counterfeits.

The referenced article suggests that Wal-Mart's prices are lower due to a shorter supply chain. Is Wal-Mart not making "enough" money? I believe they are making their targeted profit margin, as their goal is to make money for their shareholders.

Wal-Mart has done its homework at the $4 price level. It might be worth looking at the price of formulated drugs. The following is an illustrative case and has no similarity to any drug.

I have assumed an active (API) cost price is $50 per kilo. It is also assumed that the combined cost of excipients is $25 per kilo. The API to excipient ratio in the tablet is one to nine. Based on these assumptions, the component cost of the API and excipient for a 100-milligram tablet is 0.5 cent and 2.25 cents respectively. I have assumed that the cost to formulate and package is 2 cents per tablet. This brings the total cost of a finished tablet to 4.75 cents.

At $4.00 for a 30-day supply, per tablet price to the customer is 13.33 cents. Thus, there is a profit margin of 8.58 cents (about ~60%) per tablet between the formulator, distributor, and Wal-Mart. This might seem like a small number, but when you sell millions of tablets, dollars add up quickly.

The cost of the API is about 3.75% of the Wal-Mart selling price. It is a low number. The API producer has made its desired profit margin. If Wal-Mart, CVS, Target, Walgreens, and other major drug sellers, along with the API producer, want to further increase their margin, they can do that by having the API supplier and the formulator reduce their costs by implementing improved manufacturing technologies. In addition to improving their margin, the API producer and formulator will have fewer toxins to treat for safe waste disposal, resulting in reduced costs and reduced environmental impact. Besides making higher a profit through innovation, we have an obligation to our planet earth. We have to do our best to preserve its serenity and grandeur. We owe this to ourselves and for the generations to come. Thus, technology innovation is necessary.

It is conjectured that the cost of the API component of a tablet is small compared to the selling price and for that reason; there is no incentive to develop and implement better technologies. This to me is morbid thinking. It is like saying why improve our business practices if the business can make significant money because the customer would pay any price since he/she wants to live. Complacency and a lack of desire to improve invites competition and we are seeing that in the pharma world.

ENVIRONMENTALISM, TECHNOLOGY, AND HUMAN LIFE
Thursday, November 15, 2007

Environmental conservation became a way of life in the early 1970s. Every manufacturing industry had to comply with appropriate effluent standards. Conservation technologies were developed and implemented to meet necessary regulations. The chemical industry used either of the following options to comply with the regulations:

- Improve processes to maximize conversion of the raw materials to finished goods so the waste treatment load is minimized.
- Develop and use technologies to treat and dispose of the unconverted raw materials and intermediates.

Commodity, specialty, and fine chemicals, due to their competitive pressures maximized their efforts to improve the raw material conversion and relied on option number one. However, the Brahmin cousins of chemical "pharmaceuticals," that have a disease curing and life extension value due to their toxicity to bacteria, have mostly relied on the second option. This has been possible as pharmaceuticals have been able to achieve their profit margins due to relatively low competitive pressures and their ability to charge their demanded price. Customers have paid for the low conversion of raw materials in useful products, as well as disposal of the undesirable reaction byproducts.

Since 2005, the global pharmaceutical playing field has suddenly seen many players challenging their big brothers. With about $100 billion of ethical drugs coming off-patents and pressure from their generic brethren, major players are under considerable pressure to meet stock market and shareholder expectations. Theh lack of blockbuster drugs in the pipeline is also adding to their woes.

Brahmins, which until recently were "untouchables," have become vulnerable and are scrambling to retain their profit margins. It is surprising that they are following the same road that has been unsuccessfully travelled by many in the chemical and other industries. Recently we have seen layoffs, and plant closures, and they are accelerating. Research is moving offshore. I guess the Brahmins are no longer the "untouchables."

Short term with the current strategy the pharmaceuticals might be able to retain their profits. However, there is a downside and it is the disappearance of the knowledge base while the generics become stronger competitors. Generics are taking advantage of this situation and expanding as they have a growing customer base (almost everyone in the world wants lower cost drugs).

Is there a way out of this dilemma? There is and it is through manufacturing technology improvements. Companies need to develop processes where they do not have the double jeopardy, that they currently have. Double jeopardy is low raw material conversion to finished goods and then spending monies to convert the toxic materials to products, which can be properly and safely disposed of to meet the necessary environmental regulations.

Pharmaceutical manufacturing plants may meet the established environmental standards, but the small percentage of chemicals in the effluent could still be toxic to plant and aquatic life. Thus, to prevent damage to the life, pharmaceutical companies might have to meet an "ecotoxic" standard.

Ecotoxic definitions and control limits would have to be developed for most of the pharma raw materials, their intermediates, and actives. This can be a prolonged and expensive process. This will be resisted by the industry. It is difficult to conjecture the implementation costs, but they would be high. All this could raise costs and potentially make drugs more expensive.

Is there a choice and can we reduce the cost of drugs rather than raising them to achieve a certain eco-balance? Life extension also becomes a part of the economic equation. These issues can be discussed and I am not sure of the "correct" answer. However, while all this is debated, the pharmaceuticals will still be needed. Is there an economic interim solution? There is, and it is the "need to improve the active pharmaceutical manufacturing and formulation processes." Some could say that in the overall scheme of things, such an exercise is not worth the effort. Such efforts will not only reduce pharma costs but will also reduce the toxic load on the effluent wastewater treatment and soil. And finally, if someone says or believes that in the total scheme of things technology improvement and cost reductions are not important and they have no impact on our planet's environment, then we need to think about the legacy we will leave behind for the next generation and the generations to come.

Related Articles by the Author

Following is a compilation of articles that discuss various methods that can lead to manufacturing innovation. Emphasis is to cross-fertilize processes and methods from fine/specialty chemical practices to simplify the R&D methods and manufacturing technologies.

CONTINUOUS VS. BATCH MANUFACTURING

Reprinted with permission from the Paint and Coatings Industry

Emulsion and solvent-based coatings have traditionally been produced one batch at a time and packaged in appropriate containers for sale to customers. This process has been very effective. Over the years, new raw materials have been developed to enhance paint performance and lower production costs, but the method of paint manufacturing has not significantly changed.

With today's push toward lower prices and increased profits, many manufacturers are looking for ways to improve their production process. New methods such as semi-continuous and continuous manufacturing make it easier for manufacturers to increase quality and save money—benefits that might well outweigh the cost change.

Limitations of Batch Manufacturing

While advances have been made in batch manufacturing processes, there are still many limitations that affect production. New process controls are now being used to control the addition of various raw materials and to minimize the batch cycle time, but adjustments still have to be made to ensure that each batch meets its specifications. These adjustments extend batch cycle time and can lead to operating problems when the scheduled batches cannot be completed. These operating problems affect the rest of the organization, including raw-material inventory management, logistics, and sales and marketing. To reduce these problems, companies create larger than necessary buffers of raw materials and finished goods inventory, but this increases the amount of working capital required.

In batch production, small quantities of raw materials are often measured manually. This can cause batch-to-batch variation, which is accomplished by having a broad product specification that still meets market needs. The size and power differences of the plant equipment can also influence the batch. This can result in two batches of the same product that both meet specification but might be closer to the upper or lower end of the specification range. Since paint is produced in batches, the odds are that no two batches will be exactly the same. As a result, the performance of the coatings can suffer, leading to customer dissatisfaction.

Factors that can influence the cycle time and specification come from the raw materials used in batch production, including binder, pigments, and other additives, each with their own specification. If you combine all of the factors involved in paint manufacturing, it is easy to see that the industry has done a good job of producing coatings that satisfy the majority of users' needs.

A Move Toward Change

Over the past 10 to 15 years, coatings have become a commodity with growth equal to or only slightly less than the gross domestic product. Home centers have become a way of life, and they sell at "everyday low prices." This puts tremendous pressure on paint manufacturers to reduce costs in an effort to increase profits.

Increasing the profits of coatings manufacturers could require a shift in manufacturing. Other industries have changed their manufacturing processes and have improved their return on investment, but the coatings industry has lagged behind. Perhaps the time has come to implement such a change in this industry.

Semi-continuous or continuous manufacturing has been demonstrated to work and create products with better quality control, narrower specifications, and lower costs. These processes can simplify manufacturing, and improve research and development, sales and marketing, supply chain management, and IT implementation. Changing from a batch to a semi-continuous or continuous process can improve the efficiency of the raw materials used and can reduce batch cycle time, resulting in consolidation of operations. The logistics of the operation can also improve, and the supply chain management will be easier to implement.

Coatings produced by a semi-continuous or continuous process have higher consistent performance because the element of human error has been eliminated. These processes also reduce capital investment and improve asset usage due to increased throughput. Changing to these processes could also result in raw material consolidation, which will further reduce costs and improve profitability. Another benefit is lower waste generation (pollution prevention), which will also reduce the capital investment needed to treat this waste.

The concept of going from batch to continuous process is simple, but should be carefully considered because it impacts every functional group of any organization. Not every product can be produced continuously due to its total market requirement.

Conclusion

Shifting from batch manufacturing to semi-continuous and/or continuous manufacturing can provide several benefits. These include improved profitability, product quality, supply chain management and logistics, lower investment (better asset use), lower raw material costs (improved efficiency), and lower labor costs. While specific benefits can vary from company to company, it is worthwhile to take a second look at changing the manufacturing operation.

PHARMACEUTICAL PROCESSING—BATCH OR A CONTINUOUS PROCESS: A CHOICE

Reprinted with permission from Pharmaceutical Processing (www.pharmpro.com)

Most pharmaceuticals are organic or inorganic chemicals synthesized and then converted into an easily administrable dosage form for human consumption. This conversion can involve combining the active ingredient with inert materials to facilitate dispensation. Active pharmaceutical ingredients, or APIs, have to be manufactured using methods outlined in the FDA's "Guideline on General Principles of Process Validation" for safety reasons.

A review of these guidelines indicates a basic assumption that most APIs are produced by batch processes. This is probably true. The slant of these guidelines is toward batch processes rather than continuous processes. There is no mention of continuous processes in the FDA guidelines.

If the reaction chemistry and kinetics of each intermediate reaction component of an API is right (i.e., zero order or close), it is very possible to produce the API by a continuous process. One has to review the chemistry of each step and their translation to actual unit operations. Chemical engineers and chemists can easily access and demonstrate this as part of the commercialization process.

Continuous processes have the following advantages over a batch process:

- Production of a narrow specification product, i.e., higher and consistent product quality
- Reduced manufacturing cost
- Improved asset utilization
- Reduced waste.

The above advantages improve profitability. During Interphex 2004 a desire to move to continuous processes and production was expressed, but the general belief is that "If it is not broken don't fix it." In order to overcome resistance in the industry, scientists and engineers have to use proven principles to demonstrate how continuous processes can be commercialized rather than present reasons why such processes cannot be developed. Process innovation can overcome any NIH (not invented here) barrier easily.

The value and benefits of continuous processing have been discussed many times in numerous publications. In light of the benefits stated above, why is the industry still hesitant to embrace continuous processes for pharmaceutical production? I can conjecture but it may not be correct. I believe that since most chemical compounds are synthesized in the laboratory using batch processes, and their efficacy and applicability is proven under rigorous scrutiny in a batch setting, not much effort is expended to develop a continuous process. In addition, every test sample is produced by a batch process.

Once the pharmaceutical value of the API is realized, the pressure to commercialize the API is extremely high, therefore not much effort or time is expended to consider if the product can be manufactured by a continuous process. Even if the API could be produced using a continuous process, the FDA approval process, which is biased toward batch processing, may interfere and impede the development and commercialization of a continuous process. The time-intensive approval process can become a stumbling block and prevent consideration and development of improved processes. Pharmaceutical companies hesitate to spend money due to drug safety compliance requirements. An API may have a better chance of being produced by a continuous process once it is coming "off-patent."

It is my belief that among the developers of API and regulators (FDA) there are a significant number of chemists, pharmacists, and engineers who believe in continuous processing. The viability of continuous processing has to be demonstrated and it would be a win-win situation. All that is missing is for someone to step up to the plate. Recently we have been seeing higher selling prices of high volume drugs. Basic laws of economics, we all have been taught, suggest that this shouldn't happen. There are two possible explanations that may be given for the increase in selling price:

1. Increased margin.
2. Increased manufacturing costs.

The benefit of higher margins will be immediately seen in the increased profits of the pharmaceutical company. If increased profits are not seen, then the price increase can only be attributed to higher manufacturing costs, which are against the laws of economics, especially when sales are increasing. We all know that the more you produce, the more efficient one becomes and the cost is reduced. If one justifies increased costs due to meeting FDA regulations then the only remedy is to have a better batch, semi-batch, or continuous process.

The challenge is on entrepreneurs, chemists, and engineers to see which API can be produced by a better process, make it happen, and lower the healthcare costs for our fellow human beings while maintaining the profitability that is necessary for the development of better and new pharmaceuticals.

LESS IS MORE IN API PROCESS DEVELOPMENT

Reprinted with permission from Pharmaceutical Manufacturing (Pharma-Manufacturing.com)

Not only unvarying quality standards, but also profitability and elegance should drive pharmaceutical process development. Any new manufacturing process should be as profitable and simple as possible, and anything used in that process should be easy to use and execute. These rules should hold whether a manufacturing process is batch, semi-continuous, or continuous.

So why does the U.S. pharmaceutical industry persist in using complex manufacturing processes to make active pharmaceutical ingredients (APIs)? Consider patents involving API process chemistry: typically, each intermediate must be isolated, a cumbersome and costly task.

In a process involving multiple intermediates, more than one solvent is usually required, and reactions must be carried out for 24 to 48 hours or even longer. In some cases this is unavoidable, but it always introduces complexity and the need for additional steps—for example, brine washing to facilitate phase separation, the use of sulfates for drying, and vacuum stripping of solvents. All the while, reaction progress is measured by HPLC, NMR, or TLC—fine techniques all, but expensive and time consuming. This approach is fine for laboratory synthesis but can't be the rule for commercial manufacturing—at least not for any process that is to be commercially viable in today's market.

Out of necessity, API manufacturers in India and China have grasped the need for simplicity, and we've seen the results: generic therapies that cost a small fraction of the price for a nongeneric U.S. drug. While API manufacturing in the U.S. continues along this costly path, new API manufacturers in India and China are nimbly developing process chemistries that also lend themselves easily to analysis via process analytical technology (PAT) and to more advanced process control.

Simplify, Simplify

What will change this picture? The answer is simple: Less complex laboratory synthesis processes that can be scaled up easily once drug efficacy has been demonstrated. Simple processes also allow simple process control methods.

A review of pharmaceutical chemistry patents (USP 6,037,483, 6,245,913, 6,835,848, and 6,331,638) shows a "Rube Goldberg" approach, and a large number of unnecessarily complex processes. Typically, three to five steps are required to prepare each intermediate. Reactants are added over time, extending reaction time. The result: an extended time batch process.

There is nothing wrong with the chemistries themselves, but they stand in the way of process modernization and translate into higher costs to the consumer.

The following questions should guide API development efforts from their very earliest stages to ensure a final process that is as elegant, cost-efficient, and "controllable" as possible:

1. Can each reaction step be completed in minimum time? If not, how can time and costs be reduced? This is a very challenging question, but it must be addressed thoroughly, from the very first step.

2. How will reaction kinetics affect the total reaction time? Kinetics must be evaluated carefully since data will be critical to optimizing the commercial process.

3. Are you selecting the best solvents for the process? Using solvents that offer a maximum density difference between organic and aqueous phases can eliminate the need for brine washing.

4. Will the intermediate require isolation?

5. Can the same solvent be used throughout the process? This is the ideal situation, but if it's not possible, can the total number of different solvents be minimized? This will have repercussions both for solvent recovery and disposal.

6. Are you replicating commercial conditions in the development process? Can each reaction step be translated easily to an executable unit operation? Lab glassware and configurations are excellent for synthesis but don't represent reality.

7. Is the process such that it delivers quality product rather than quality is achieved by testing the product?

Of course, laboratory methods can't fully simulate plant conditions. Sophisticated control methods aren't practical for use in the lab. Nevertheless, there are steps that chemists and engineers can take to apply the best available technology and methods.

Economical Choices

As an example, consider amine diazotization, which is typically conducted at low temperatures with ice—a safe but inefficient method that leads to disposal issues. This reaction could instead be conducted safely at room temperature.

Let us analyze each alternative: first, the traditional method, in which an amine is added to water, and sodium nitrite and a mineral acid are added to complete the reaction at temperatures of 0°C or lower. The ice and excess water are needed to control this exothermic reaction.

But consider the historical reasons for using this process. When this reaction was first developed, ice and excess water were the best and cheapest heat sink available. The plants practicing these technologies were usually located near rivers where plenty of free cold water was available. The effluent water after use was treated with the best available technology and sent back to the river.

Now let's consider the alternative: conducting this reaction at room temperature. This approach could be safely taken if one follows the reaction chemistry along with physical chemistry. A back mix type reactor with stoichiometry controls can be used to control the exotherm and conduct the reaction safely. Although an alternative is possible, most recent patents still describe the older method.

But sometimes the older method can be the most economical. Consider the choice between chemical reduction, typically used to convert a nitro to an amino group, and catalytic hydrogenation. Chemical reduction leads to byproducts that must be treated before disposal, and it's generally a messy process. Catalytic hydrogenation is much cleaner and is now the method of choice in developed countries.

However, in developing nations, where catalytic hydrogenation was too expensive, manufacturers have instead optimized chemical reduction—and they've recovered, purified, and sold byproducts, leading to processes that are more economical.

Let's review some recent API process patents, and the potential cost-reduction opportunities for each.

U.S. Patent # 6,037,483

This patent describes preparation of 3-Bromomethyl-3-methyloxetane. The process is complex: Preparation of oxetane is completed over two days and involves isolation of an intermediate and use of a solvent. A close examination of chemistry and reaction kinetics suggests that it is possible to conduct the reaction in about three hours without the use of the solvent and isolation of the intermediate. The yield is over 85% and productivity much higher compared to the process described. If the process as described in the patent is commercialized, it will definitely be expensive.

U.S. Patent #6,245,913

This is the synthetic procedure for 5-methoxy-2-[(4-methoxy-3,5-dimethyl-2-pyridinyl)-methylthio]-IH-benzimidazole hydrochloride and its conversion to omeprazole.

This patent has simplified the process from previous chemistries, but, still, 11 major steps are needed to produce the product. There are isolation and purification steps in each of the 11 steps. Organic chemists and engineers are very creative in minimizing and/or eliminating isolation. Their creativity has to be tested as it will lead to a simpler process that can be controlled to minimize reaction time. Different solvents are used in the process, adding to the overall complexity. I believe that there are simplification opportunities for each intermediate product on commercial scale. These opportunities come from reaction time and work up procedures. Is the isolation of the intermediates necessary? My estimate is that the overall process yield from the starting raw material is about 20–25%. This presents a significant opportunity to increase yield, which can add substantial savings to the bottom line.

U.S. Patent #6,867,306

The present invention discusses a novel process for the synthesis of [R-(R*,R*)]-2-(4-fluorophenyl)-B,D-dihydroxy-5-(1-methylethyl)-3-phe-

nyl-4- [(phenylamino)carbonyl]-1H-pyrrole-1-heptanoic acid hemi calcium, atorvastatin.

This patent relates to statin synthesis and claims to be "a new, improved, economical, and commercially feasible method." The patent has
nine major steps. Within each step, there are four to ten steps. These
add to the processing complexity and cGMP becomes an operating challenge. It is my conjecture that the process as described if commercialized
would take about one week to make a batch.

Compare these economics to those of the Chinese fermentation process. It is necessary to simplify the process to minimal steps so that isolation, if necessary, is easy and the product cost is significantly less than
the comparable product made anywhere in the world. (Reference: UBS
Investment Research, October 8, 2004.)

U.S. Patent # 6,835,848

The present invention relates to a new and industrially advantageous
one-pot process for the preparation of alkyl 3-cyclopropyl amino-2-[2,4-
dibromo-3-(difluoromethoxy) benzoyl]-2-propenoates, in which R represents methyl or ethyl, which are valuable intermediates for the production of highly active antibacterial quinolone medicaments.

The title of the patent calls for "one pot" synthesis. However, six steps
are carried out in that pot over two days. Although they are all simple
steps and could lead to a manufacturing-friendly process, it could be
possible to shorten the time required significantly, and even move from
batch to continuous production. The process chemistry as it is described
cannot be completed in one pot unless that pot is used repeatedly. It
needs simplification to be a simple process.

U.S. Patent # 6,331,638

A process for the preparation of the esters of 1,8-disubstituted-1,3,4,9-
tetrahydropyrano (3,4-b)-indole-1-acetic acid is delineated.

Although this seems simple, synthesis takes about 10 to 12 hours to
complete the reaction. A batch process time of one to two hours or a
continuous process would make the total process economics extremely
promising. I believe such time reduction may be possible for this process.
These may be possible by a careful review of the reaction stoichiometry
and kinetics.

Reaction completion and purity are measured using NMR, HPLC, TLC,
or other sophisticated analytical methods. These prolong the cycle time.
For commercial operation, simple test methods that give the desired test
result would be helpful.

To reiterate, anyone embarking on API process development needs to ask some hard-hitting questions as early as possible in the process, and to ensure that they are answered. In a nutshell, "Is this process as inexpensive, profitable, safe and environmentally friendly as it can be?"

Development specialists need to put themselves in operators' shoes: Ask yourselves, "If I had to operate the process in a manufacturing environment, how would I feel? Is this really the best that I can do?"

This approach will simplify process control, via PAT, eliminate rework, slash inventories, and result in "right first time" product quality.

As in art, in process development, less is more. In today's global economy, U.S. scientists and engineers are competing with the best chemists and engineers of every country in the world, whose work is already guided by this principle.

Cost efficient methods will allow manufacturers to embrace more advanced process control, facilitating cGMP compliance and, perhaps most importantly, reducing product costs to the ultimate consumer.

API MANUFACTURE—SIMPLIFICATION AND PAT

Reprinted with permission from Pharmaceutical Processing
(www.pharmpro.com)

Worldwide approval, safety, and efficacy of administered drugs are controlled by government agencies similar to the Food and Drug Administration (FDA) in the United States. These agencies also make sure that the manufacturing processes produce products of consistent quality and are independent of the day or the year of manufacture.

Since 2001, FDA's Center for Drug Evaluation and Research (CDER) has been encouraging pharmaceutical companies to apply modern process analytical technologies (PAT) to the manufacture of drugs. CDER anticipates that application of PAT not only will improve product quality but also will lead to process innovations, thereby reducing the costs of drugs.

However, based on many published articles one gets the impression that manufacturing innovation processes at pharmaceutical companies are still lagging. The reasons for lack of change are very understandable. No one wants to change the drug manufacturing process when it is in clinical trials and revalidation can be expensive. However, process innovation should be included in the development and commercialization of new drugs. As process innovation benefits are realized with new drugs, improvement will be adopted across the board.

Those involved with drug development are worried that the costs associated with regulatory filings due to process innovation will be high. The actual reason is not cost but the reluctance to change. Change is difficult. Irrespective of FDA rules and practices, companies can achieve their profit margins comfortably until the patent's protection runs out. Thus, they may not see any need to change. After the patent protection runs out, generic producers take over the market. Generic drug manufacturers follow practices similar to those found at the companies who initially developed the drug.

It has been suggested that drug manufacturing is an art. It has also been suggested that manufacturing in pharmaceutical companies is treated as an orphan and/or a necessary evil. In fact, drug manufacturing is an extremely important element of the total system and needs attention. Drug manufacturing, especially API manufacture, is not an art, but the production of specialty/fine chemicals that have pharmaceutical value. Principles of chemical kinetics, process control, and unit operations are part of chemical engineering curriculum and are being applied in many chemical and petrochemical manufacturing processes. Thus, process controls and PAT can also be applied to the manufacture of pharmaceutical chemicals. It is very possible that "quality by design" (QBD) is being considered right now by various companies. However, the FDA would like to see its implementation hastened.

In its 2004 report,[1] the FDA clearly explains the "current state" and the "desired state" of manufacturing. An excerpt of the "current state" from this report is:

"Pharmaceutical manufacturing operations are inefficient and costly. Compared to other industrial sectors, the rate of introduction of modern engineering process design principles, new measurement and control technologies, and knowledge management systems is low. Opportunities for improving efficiency and quality assurance through an improved focus on design and control, from an engineering perspective, are not generally well recognized. For example, when discussions at the FDA Science Board and Advisory Committee for Pharmaceutical Science shed light on the current low efficiency and its cost implications (e.g., costs associated with manufacturing can far exceed those for research and development operations in innovator pharmaceutical firms) many at FDA had difficulty understanding this and common reactions were 'how could this be possible?' or 'this can't be true.' Regulators and many in manufacturing operations express their frustration by suggesting that manufacturing is a 'step-child' in this industry, and that there is no economic motivation (e.g., cost and price difference) for improvement. Other suggestions include a general lack of systems perspective, organizational barriers that inhibit exchange of

knowledge, and the attitude that much of pharmaceutical formulation and process development is an 'art.' Some in pharmaceutical development suggest that there are very limited opportunities ('development time crunch') to realize and/ or demonstrate the level of science underlying current formulation and process development efforts."

An excerpt of the "desired state" is:

"Improving the foundation of manufacturing science in our current manufacturing practices should be the primary basis for moving away from the corrective action 'crisis' to continuous improvement. The 'desired state' for pharmaceutical manufacturing in the 21st century therefore emphasizes and aims to improve knowledge on design and understanding of product and processes."

The FDA wants the pharmaceutical industry to implement modern process control technologies. Dr. Ajaz Hussain (FDA) in a recent article states, "Process understanding can be a foundation for innovation and continuous improvement in pharmaceutical development and manufacturing."

Manufacturing has to move from "quality by inspection" to "quality by design." Unless we are able to achieve the QBD goal, manufacturing will depend on inspection and that method is always expensive. This article suggests some methods that can simplify manufacture of active pharmaceutical ingredients (API) and lead to "quality by design."

Even after the FDA's initiative of "the desired state," the perceived resistance to change may be due to industry's inability to apply QBD to the entire drug manufacturing process, which includes API manufacture and dosage manufacture. The majority of published articles on this subject relate to the challenges faced in the manufacture of single dose. Issues related to API manufacturing are seldom reviewed. It is important that issues related to API manufacturing be addressed, as without an API, we do not have a drug. It may be easier if the industry undertakes simplification of API manufacturing and dose manufacturing separately.

Table B.1 illustrates the financial impact of a 1000 pound API batch that has to be dumped. Using a selling price of $0.50 per dose, dumping could result in revenue loss of over $2.5 million (this includes the disposal cost) and PBT loss of about $0.5 million. Thus, it is in the best interest of pharmaceutical companies to produce quality products all the time. If an API has to be reworked to meet quality or is recalled due to quality control problems, it will cost money and reduce after tax profits.

Obviously, the loss of a single API pound can be expensive to the company both at the revenue line and the profit after tax line, and must be avoided. If API process development is guided by the following considerations, it is very possible to eliminate such losses. These are age old and time proven methods but need to be applied thoughtfully to the pharmaceutical manufacturing processes.

1. Generally known principles of chemistry, physics, and engineering have to be applied to the process synthesis and development from the very first experiment. Chemical kinetics determine the rate of reaction and conversion. Reaction rates and stoichiometry have to be such that they give the highest yield in the shortest time. This is necessary and results in consistent and high quality product.

2. Chemistry and process (reaction, separation, and purification) in the laboratory have to be simple so that they can be easily translated to a commercial operation.

3. Is the intermediate separation and purification step necessary?

4. Process has to be such that its environmental impact on the total plant, including waste disposal, is minimum.

5. Is it possible to minimize the number of solvents in the whole process? Since solvents have to be recovered or disposed, they add to the cost and processing time if many different solvents are used in the process.

6. Conversion measurement methods have to be simple and such that they can be easily applied in a manufacturing environment. Use of exotic and expensive analytical instruments is acceptable in the laboratory, but it has to be minimized for a commercial operation as they prolong manufacturing processes. Simplicity is needed.

7. Process has to produce quality first time and all the time. Rework and recall are not an option.

8. Since most APIs are produced using batch processes, the on stream time (OST), (defined as actual run time divided by available time for process equipment associated with the process) for such operations would be 60% or lower. A productive operation should have OST in the high 80–90% range. OST inefficiencies are compensated by having an operation of higher than needed capacity, i.e., over-investment.

Table B.1 The financial impact of loss of an API batch.

API batch size, pounds	1000
Dosage, mg	100
Total dosage from the batch	4,530,000
Selling price of each dose, $	0.50
Lost Revenue, $	2,265,000
Profit before tax	20%
Lost profit, $	453,000

The above may seem very demanding for a laboratory development process, but are essential for a manufacturing environment. Again, process commercialization starts with the laboratory process chemistry and development, not after the pharmaceutical efficacy of a specialty chemical has been proven. Implementation of these practices will reduce time to market, ease validation, simplify commercialization, and reduce manufacturing costs. It is well known that if an API shows promise, commercialization becomes paramount and there is no time to optimize the manufacturing process. Time, effort and cost can become a deterrent if an API is produced by a non-optimized process and the drug is in Phase II studies. The pharmaceutical industry is operating under these conditions today.

The application of exotic analytical instruments and spectroscopic methods is necessary in the laboratory, but they become a bottleneck in the manufacturing environment. Their use in a manufacturing environment has to be very judicious. We have to adopt methods that are economical and give us the desired information, but not at the expense of processing time. Cost value justification has to be considered at every step.

If methodologies similar to those mentioned above are used in the development of APIs, it is very possible that commercially available process controls can be easily applied and the industry will be on the path of "quality by design" and move away from "quality by inspection." The process will then deliver quality product all the time in a manufacturing environment. Simplicity is the key to have a low investment and simple operating process. If the process chemistry and its execution can be simplified, it is possible to have a semi-continuous or a continuous manufacturing process[2] for high volume products.

An argument can be presented that none of the above is necessary during the initial screening and development. Pharmaceutical companies have to adopt a new premise that every specialty chemical synthesized has a pharmaceutical value and will become a high revenue product. Thus, there is no time to come back to the laboratory and optimize the synthesis, chemistry, and purification as product commercialization ahead of competition becomes the primary goal.

Following are some observations from a few patents. These are a reflection of the process development methodology in the laboratory.

1. Phase separation and intermediate purification: patents include multiple brine washes, solvent extractions, and drying using magnesium sulphate to isolate intermediates. Are these steps necessary? My conjecture is that these steps are being used to produce a pure intermediate due to low reaction conversion. An attempt has to be made to achieve higher product quality through higher con-

version rather than by multi-step purification, which costs time and money.

2. Multiple solvents are used. Are they necessary? How can one minimize the number of solvents used in a process and at a site? Many of the solvents are high priced research chemicals. Can commercially available, i.e., low cost solvents be used instead?

3. Stoichiometry and reaction conditions have to be such that they result in the highest yield in the least time. If an API has six steps with 98% yield for each step, the overall yield would be about 88.5%. If the yield drops from 98% to 95% (considered good), the overall yield drops to 73.5% (not considered good and needs improvement). (At least that is what we were taught and practiced.)

4. Reaction times exceed 8–24 hours. This is expensive.

It is easy to explain the reasons and rationale for the above practices. Methods to improve these practices are known and need to be implemented. We have to emphasize that every specialty/fine chemical is an API, and the manufacturing process has to be simple. If we choose this path, the process of continuous improvement will become the rule instead of the exception.

In the chemical industry it is well known that every specialty chemical (APIs produced by pharmaceutical companies) will have a high profit margin and as soon as it becomes a commodity (a generic produced by generic pharmaceutical companies) profit margins drop. Thus, if an ethical drug is produced with the best process, its cost will be minimal. This will also deter generic producers from taking over after patent expiration, as it will be a bigger challenge for them to improve the process technology. Under the current scenario, they are very successfully picking the low hanging fruit of the big pharmaceutical companies.

It is well recognized that patents as written are not how the chemistries are commercially practiced. However, they show opportunities. The following patents are cases in point.

USP 6,747,153 for Tiotropium Bromide to Boehringer Ingelheim based on the chemistry described could take over 200 hours to complete a batch. If all of the steps (15+) could be finished in less than 100 hours, which is a long time to make a batch, it would still be progress. The process yield is about 48.2%. This is a low yield and presents significant process improvement opportunities.

Lipitor patents 4,681,893 (organic synthesis) and 5,273,995 (chiral synthesis) have many steps. Either process as described in the patent suggests separation and isolation using multiple brine, solvent washings, and magnesium sulfate drying to isolate intermediates before the final product is produced. Kinetics of the reaction steps can be improved to reduce batch cycle time. There are opportunities here as well.

Naproxen patent 4,623,736 has about six steps. This is a high volume product. It is possible to convert this batch process to a semi-continuous/ continuous process using a stepwise approach.

Significant laboratory effort is put into developing methods and procedures that are used in writing a patent. If the same effort is put into optimizing the process in the laboratory, it is possible that one can have a simple process. Since the FDA has suggested a path for the industry to follow, it is in everyone's best interest that we collectively develop and implement a plan to achieve the desired state. It will be a win-win situation. The economic benefits are there, the industry just needs to capitalize on them.

References

1. Innovation and Continuous Improvement in Pharmaceutical Manufacturing. The PAT Team and Manufacturing Science Working Group Report. *http://www.fda.gov/cder/gmp/gmp2004/manufSciWP.pdf.*
2. Malhotra, Girish. "Pharmaceutical Processing—Batch or a Continuous Process: A Choice." *Pharmaceutical Processing.* March 2005;16.

QUALITY BY DESIGN (QBD): MYTH OR REALITY?

Reprinted with permission from Pharmaceutical Processing (www.pharmpro.com)

In the United States, the FDA's initiative on nudging the pharmaceutical industry to invent, develop, and commercialize products using technologies that will result in product quality by design (QBD) is a challenging task. It is also a noble task that will have major business process implications and ultimately high financial impact on healthcare costs.

Through presentations and pilot programs, the FDA is making its case to move the industry toward a win-win situation for all involved, especially consumers. These outlined items, if submitted, would assist in the approval process and allow continuous improvements. Since this is uncharted territory for drug developers and reviewers, it is necessary for the developers to present information that will convince the reviewers that the process will produce QBD, the "desired state," rather than quality by inspection (QBI), the "present state." Because this is a change in process and mindset, there will naturally be apprehension on the part of industry as it has traditionally worked in a defined comfort zone and is not sure about the value of change.

QBI versus QBD

With that said, processes are developed and commercialized by chemists and engineers with the involvement of regulatory personnel to comply with the FDA, EPA, and OSHA rules and regulations. In this evolution, one has to optimize the impact of their actions when it comes to the total business process. The current QBI methodology creates business processes and inventories that need infrastructure to support quality pharmaceutical production. This costs money and the consumers pay for it. QBD can be equated to just in time (JIT) as the industry will have total control of the manufacturing process. Volumes have been written about the value of JIT and we all know that it smoothes out the total business process.

The FDA in its expectation of process analytical technology (PAT) framework and QBD stipulate a complete understanding of the interaction of raw materials and intermediates and control of process parameters. In a commercial operation, this will result in a QBD rather than QBI product. This is easier said than done. It requires combined application of knowledge, common sense and stepping out of bounds to establish a new paradigm.

Improving the Manufacturing Process

In this paper, I am reviewing a chemistry outlined in a patent and sharing some of the opportunities to improve the manufacturing process. Process controls can be applied which will produce consistent quality product with no or minimal in-process testing. Similar methods and observations can be incorporated in the development of new chemistries so that we have a QBD process from the outset.

I have chosen production of an active ingredient rather than the formulation of the active pharmaceutical ingredient (API) with excipients. The reason for choosing reaction chemistry is its complexity and the length of time it takes to produce an API. In addition, much has been written about the formulation aspects and very little to none has been written about the manufacture of APIs. Thus, it needs proper consideration.

For the development of a process and chemistry of APIs an alternate approach could be considered. This is "out the box" thinking but worth consideration. Instead of having a mindset that we are developing a "pharmaceutical," it might be easier to consider that we are developing a "specialty chemical" that might have pharmaceutical value. This should simplify many of the drug development processes.

In this alternate approach, once an optimum process has been developed for a specialty chemical where we know the interaction of each raw material and intermediates, the critical parameters and how to control them, every regulatory requirement can be included to meet the necessary standards. Since we have to apply regulatory requirements only on one process, the product and process development is simplified. Such an approach might also be a way to reduce the "time to market." I believe that if we are able meet quality and performance specifications all the time, we might have to contend with fewer regulations also.

By using the "right" process we will produce a "quality" product, which in turn will reduce waste, work in process inventories, regulatory oversight and bureaucracy, i.e., simplify the business process. Doing it right the first time by a repeatable process is the key and has to be the method of choice. If we are able to accomplish this from the outset, we would not have to live with 2.5 sigma processes, and the process of continuous improvement would be less expensive compared to after the fact improvements.

Process Considerations

The following considerations are necessary in the development, simplification and commercialization of the "right" process. Most of these are being used in the manufacture of chemicals. The understanding and application of these also allows the control of processes using commercially available process control technologies. The following are taught in chemical engineering curriculum, thus these are not new.

1. Total process feasibility. Each unit process step has to be reviewed individually and collectively.
2. Is the stoichiometry optimized?
3. Are the heat and mass balance optimized?
4. Are the reaction kinetics understood and applied to simplify the process?
5. Are proper unit operations being used?
6. Are the necessary steps in place to reduce the cycle time?
7. Can a single solvent be used for the whole process? This economizes solvent recovery and the related investment.
8. Can we eliminate isolation of intermediates?
9. Are the raw materials to be used easy to handle?
10. How can the phase separation be improved and simplified, if it is part of the process?
11. How can I improve the conversion of each process step? Lower conversion means that there is raw material loss, which has to be recovered and/or treated in the effluent system or disposed of as hazardous waste. Lower conversion also means that unless the

unconverted raw materials or impurities are removed prior to the subsequent reaction steps, additional impurities will be created, adding to the process complexity.

12. Are the safety requirements met and is the process safe?

13. If the developers were operating the process, what process modifications and/or additions would be included to have the simplest process?

14. Is the process meeting all of the environmental standards?

15. Is the rework eliminated and/or minimized?

16. Is the process economical? A thorough understanding of every interaction allows one to have a complete grasp of the impact of every process change and its influence on the product quality. If the above considerations are followed all the time QBD becomes a natural part of the development process. In addition to the above considerations, each chemistry and process has its nuances and if recognized and implemented can simplify the manufacturing processes further. The incorporation also allows one to have complete control of the process. One can react to any unexpected changes and deliver quality. From my experiences, it also allows one to repeat the mistakes. If this can be done, developers will have done an excellent job. It would be like driving a car smoothly under every condition. Today's pharmaceutical manufacturing can be compared to an automobile driving us vs. us driving the automobile.

The application of the above considerations is part of the reviewed process. It is also possible that some of the process observations mentioned below, if implemented, can lead to a continuous process. It is well known that a continuous process is more economical than a batch process and produces consistent quality, i.e., QBD.

Dr. Moheb Nasr, director, CDER's Office of New Drug Quality Assessment, Dr. Janet Woodcock, Dr. Scott Gottlieb, and others at the FDA have echoed the sentiment well known in the chemical industry that "the pharmaceutical industry can only realize the full benefit of QBD by developing and implementing continuous processing."

U.S. Patent 4,623,736

This patent covers the synthesis of ibuprofen and naproxen. The patent describes different routes. However, I have only reviewed the preparation of ibuprofen from example 1. Similar observations can be applied for the preparation of naproxen (example 2) also.

Process description: Isobutyl benzene is reacted with alpha-chloropropionyl chloride and aluminum chloride with methylene chloride as a solvent. The reaction is carried out at 0 to -5°C. Excess aluminum chloride is neutralized with dilute hydrochloric acid. Organic and aqueous phases are separated. The organics in the water phase are extracted with methylene chloride. The combined organic phase is washed with sodium bicarbonate to pH 7–8. The organic phase is heated to remove methylene chloride and heptane is added. The resulting ketone is then reacted with neopentyl glycol in presence of concentrated sulfuric acid. The material is refluxed at about 97–107°C. Water is azeotroped. After the reaction is complete, about eight hours, the reaction mass is washed with a diluted solution of sodium bicarbonate. All of the organic phases are combined and washed with water. Water and organic phases are separated. Heptane is removed under vacuum and ketal oil is obtained.

Ketal oil is heated to about 140°C and this removes any remaining heptane. A catalytic amount of zinc 2-ethylhexanoate dissolved in heptane is slowly added and the temperature is maintained at 140–150°C. Since the reaction temperature is above the boiling point of heptane, all the heptane boils off. Upon reaction completion, the liquid is cooled to about 25°C and a filter-aid is added to absorb the zinc salt. Zinc salt is filtered. Liquid is carbon treated and filtered. Heptane used for solid washing is combined with the process liquid.

Chloroester produced above is heated to 95–100°C and sodium hydroxide solution is added. The reaction mass is heated to about 95°C. Once the hydrolysis is complete, water is added to dissolve the sodium salt in water. Aqueous solution is cooled to 0°C and crystals of sodium ibuprofen are filtered, washed with heptane, and dried.

Sodium salt is converted to the acid by dissolving it in water and adding of heptane and hydrochloric acid. Some of the heptane is distilled and the acid is crystallized from heptane.

Review: The above synthesis is classical chemistry and if one did not know that the final product 2-[4-(2-methylpropyl) phenyl] propanoic acid has a pharmaceutical value, it would be treated as a specialty chemical synthesis. Since it is a pharmaceutical, it needs to meet all of the requirements for human consumption. We have to keep in mind that most of the patents are based on laboratory procedures and for commercial production, we need to have a viable and economic process. Thus, from a business perspective the process might need modifications to be an economic commercial operation. Considerations mentioned above should be applied to each step individually and collectively to the whole process.

Step 1: In the first step of the reaction, there are few opportunities. The Friedel-Crafts reaction is carried out with methylene chloride as the solvent. After neutralization, a new solvent (heptane) is added and it is used in the rest of the process. If we can substitute heptane for methylene chloride from the beginning, only one solvent will be used. This reduces the investment necessary for methylene chloride tank, recovery, and handling. Use of heptane can lead to a price advantage due to purchase of a larger quantity of single solvent.

Solid handling of aluminum chloride requires special methods. Aluminum chloride can be slurried in heptane, fed as a liquid, and this can simplify processing and handling. Reaction is carried out at 0 to -5°C. Most of the Friedel Crafts reactions are zero order reactions. To control the exotherm, the process as described suggests the slow addition of propionyl chloride. In addition, a proper feeding system with feed, temperature, and other controllers would be needed to control the kinetics and temperature. It is well known that if a reaction can be conducted at higher temperatures, the reaction time can be reduced. With a properly designed system, it is possible to have a continuous process for this step. (This could be labeled unsafe but since I have practiced it I can write about it.) Commercially available process controllers are used on a routine basis. A properly designed reaction control system will have high conversion and might not require in-process testing of conversion. In a continuous process, phases can be separated continuously without any interface controls. All these improve productivity.

The process as suggested uses about 9% excess of alpha-chloropropionyl chloride and 28% excess of aluminum chloride. If the raw material use can be optimized it will reduce chemicals needed for neutralization. This will also reduce raw material and waste treatment costs and improve productivity.

Step 2: In the second step, product from step 1 is reacted with neopentyl glycol to produce a ketal. The mixture is heated to about 90°C and the condensation reaction is carried in the presence of concentrated sulfuric acid. The mixture is heated to 97–107°C and the produced water is azeotroped. The reaction is complete in about eight hours. The mixture is cooled and neutralized with sodium bicarbonate solution. Each phase is washed and heptane is removed from the organic phase under vacuum. Glycol could be added as a molten liquid. With proper heat and mass balance, the condensation can be accelerated and controlled, acid neutralized, and heptane removed. It is possible to do the condensation continuously with proper control of glycol addition. The process as suggested uses about 39% excess glycol based on isobutyl benzene. There is an opportunity to reduce this amount and therefore lower cost.

Step 3: Ketal from step 2 is converted to ibuprofen ester using zinc 2-ethylhexanoate. The reaction is carried out at about 135–150°C. Rearrangement reaction is complete in about two hours. The reaction mass is cooled to 20–30°C and a filter aid is added. Heptane is added and the solids filtered out. Filtered liquid is carbon treated and filtered. Since ketal and the catalyst are liquids, processing and controls can be done by classical methods. It is possible that the reaction time can be reduced significantly to have a continuous process that would include carbon treatment and filtration.

Steps 4 and 5: These are classical cases of producing a sodium salt through crystallization and converting the sodium salt to its acid, crystallization, filtration, and drying. Steps 4 and 5 are routine unit processes and operations that can be done continuously with proper process controls to produce a product that meets quality specifications all the time.

Possible Additional Improvements: Once the steps of a continuous ibuprofen process are assembled, it might be possible to achieve additional improvements, i.e., yield improvement due to changes in addition methods and operating conditions. This is all possible, since the developers have done a good job of keeping the chemistry and processing of ibuprofen and naproxen simple. Again, in order to improve productivity, cost and achieve quality, developers, chemists, and engineers have to review each reaction step individually and collectively and simplify them to have the most productive process. Mr. Thomas E. Burakowski from Boehringer-Ingelheim Pharmaceuticals, Inc. in a recent article mentions that in the manufacture of pharmaceuticals it is not uncommon to have 20–30 synthesis steps. If the yield in each of the 20 steps is 98% (highly unlikely), the overall process yield would be about 66.8%. I believe processes with this many steps would have a long cycle time and would be uneconomical in the specialty chemical world. Since we need to produce the API, we have opportunities. Table B.2 shows the overall yield of processes that have 5 or 10 steps with 95, 80, or 70% yield for each reaction step. Yields are based on the key starting material. Lower yields suggest cost reduction, productivity improvement, and quality enhancement opportunities. Every effort should be made to minimize the yield loss. Higher yield has to be the goal of every process. In the pharma world, we

Table B.2 The effect of the number of steps and yield.

Overall yield, %			
Yield per step, %	95	80	70
Number of Steps			
5	77.4	32.7	16.8
10	59.9	10.7	2.8

are able to recover such losses through the price of the drug since people will pay to extend their life. However, in the pure chemical world such processes would be considered uneconomical.

We have acknowledged "quality by inspection" as an acceptable business model for life-prolonging drugs and companies have been able to achieve their profit margins. Thus, there is very little incentive to optimize drug manufacturing processes. The FDA has echoed this in its assessment of the current manufacturing technology.[1] Lack of optimization results in variable product quality. We are able to meet stringent quality standards through isolation, purification, and checks at every step. This is very similar to steps that are taken in a laboratory or pilot plant to develop theses products. One could call pharma manufacturing a large-scale laboratory or pilot plant. Since we have to inspect quality, the whole business process is burdened with unproductive costs.

There are opportunities in every synthesis and one has to capitalize on them from the outset. It is well recognized that there is time pressure for first to market, but lack of an optimum process burdens the business process and no one has ever calculated the economic impact of this burden. Lowering or removing these burdens will lower costs and improve profits further.

A brief review of several patents suggests simplification opportunities.[2] A review of Ganciclovir intermediate manufacture[3] suggests that the synthesis can be modified and simplified to produce only the desired isomer only. This can create a new process, which could out the bounds of these patents, and QBD product. Manufacture of Modafinil (Provigil)[4] for the treatment of narcolepsy can be simplified to a continuous process. Similarly, Omeprazole® (Prilosec) could be produced using a continuous process.

A complete knowledge of every interaction of raw materials and process conditions will produce quality with very few or no isolation of intermediates and in-process tests, a real QBD. Thus, QBD is not a myth. It is feasible and achievable.

The drive to achieve QBD for any API being developed in the laboratory should not start or end at the understanding of the interaction of raw materials, their reaction conditions, and physical properties for chemicals that are used to produce any product. It can also be applied to the existing APIs as it will simplify the total business process and improve profitability.

It may be time to explore alternate ways to simplify process development and manufacturing methods for APIs so that after drug efficacy, "first time quality" becomes a driving force in pharma manufacturing. Any company achieving QBD will have the optimum process (i.e., lowest cost) for the used chemistry. This can act as a deterrent to the entry of generic producers. In addition, it will keep the profitability at a significantly high level beyond the patent life compared to the current scenario.

References

1. Innovation and Continuous Improvement in Pharmaceutical Manufacturing. The PAT Team and Manufacturing Science Working Group Report. *http://www.fda.gov/cder/gmp/gmp2004/manufSciWP.pdf.*
2. Malhotra, Girish. "API Manufacture—Simplification and PAT." *Pharmaceutical Processing.* November 1, 2005;24.
3. USP 5,821,367, USP 5,565,565, USP 6,043,364, USP 7078524, USP publication 2005/0176956, and USP publication 2006/0142574.
4. USP 6,875,893.
5. USP 5,958,955 and US 2006/0084811 A1.

BIG PHARMA: WHO'S YOUR ROLE MODEL, TOYOTA OR EDSEL?

Reprinted with permission from Pharmaceutical Manufacturing (PharmaManufacturing.com)

Generic drug manufacturers must grapple with the tightest of all pharmaceutical profit margins. This is especially true of generic manufacturers in developing countries. Branded drug makers had not suffered from such constraints in the past, but this situation is about to change.

As key drug patents expire over the next five years, manufacturers of name-brand pharmaceuticals will lose about $100 billion in revenue and associated profit to generics manufacturers, at a time when there are fewer blockbuster drug candidates in the pipeline and costs for drug discovery, development, and commercialization have escalated.

Branded pharma is also seeing increased competition from generic manufacturers in developing nations. Despite cost and other pressures, Indian and Chinese generic manufacturers have challenged existing business models, and are realizing the benefits of intellectual property (IP) protection by developing patents of their own, channeling increased profits into developing branded drugs.

Big Pharma could learn a few lessons in manufacturing efficiency and agility from some of its generic competitors in Asia. History is full of examples like this: 60 years ago, an obscure company in Japan began to pursue manufacturing innovation. Toyota soon surprised Detroit and the world with what it had learned.

Today, name-brand drug companies are attempting to improve margins by consolidating or outsourcing manufacturing, repositioning their R&D, and realigning their sales forces. But have enough of these companies looked at how they are designing their processes for API manufacturing? So far, unlike generics manufacturers, especially those operating in the developing world, branded drug makers have had little incentive to reduce drug prices to consumers, or to reduce their own costs.

Efforts to cut manufacturing costs must be escalated. Today, most drugs are still produced by inherently inefficient and costly batch processes. Very little has been done to advance improved batch and continuous processes in pharmaceutical manufacturing. This is a real shame because the benefits of better manufacturing methods and technologies can be realized quickly.

The FDA is encouraging implementation of process improvements in the manufacture of active pharmaceutical ingredients (API) and dosage formulations. Conversion yields of most existing commercial API processes are poor, resulting in wasted raw materials, inefficient manufacturing, and unnecessarily high costs to the consumer. A quick and concerted effort is needed to improve API manufacturing, involving a thorough understanding of process chemistry, reaction kinetics, physical chemistry, and unit processes and their incorporation into the simplest unit operations. The results will not only be high conversion yield but "greener" manufacturing processes.

As to the naysayers, let history be their guide. In the early 1970s, when the global environmental movement arose, many believed that effluent regulations would bring manufacturing to a stop. They didn't. Instead they improved efficiency, as those who complied with the new regulations improved their processes, which led to greater profitability.

Pharmaceutical manufacturing has been meeting environmental standards and FDA guidelines. But FDA and EPA cannot force manufacturers to improve their operations or reduce costs. Change must come from within.

Will that change ever come? Recently, I saw a reaction process that took about 60 hours to carry out. The same reaction step could be done in a different way without altering the basic process, reducing the cycle time by 75% and increasing the process capacity by the same percentage. ROI for this project could be realized in about three months.

There are many other examples of different API chemistries where improving cycle times and conversion yield will improve margins, reduce waste and environmental emissions, and reduce the cost to the customer—a win-win for all.

Toyota transformed automaking, not only with its business process management system, but also with the engineering and improved fuel efficiency of its engines. If big pharma does not begin to innovate in its manufacturing and use of process control, it, too, may meet the fate of automakers who were slow to respond to market forces.

PHARMACEUTICALS, THEIR MANUFACTURING METHODS, ECOTOXICOLOGY, AND HUMAN LIFE RELATIONSHIP

Reprinted with permission from Pharmaceutical Processing (www.pharmpro.com)

A recent study[1] by Dr. Joakim Larsson of The Sahlgrenska Academy at Göteborg University, of the effluent water from the Patancheru wastewater treatment facility in Hyderabad, India is going to cause a global uproar about environmental regulations and pharmaceutical and chemical operations, if it has not already done so.[2] In and around Hyderabad, a who's who of the Indian pharmaceutical and active pharmaceutical ingredient producers and formulators have their plants. This study gives us a snapshot of the wastewater treatment plant operated by Patancheru Enviro-Tech Limited (PETL), its operating efficiency, and allows us to conjecture about the state of affairs around the chemical and pharmaceutical plants. In his recent paper, Larsson speculates on the "impact of effluents" from active pharmaceutical plants worldwide. Every active pharmaceutical ingredient (API) and drug (combination of chemicals) are chemicals. They all have disease curing (toxic to bacteria) and life extending value. Effluent waters of API and drug manufacturing facilities have residues of these chemicals. Their levels can be toxic to the aquatic and soil life and have broad implications. Larsson notes that the cumulative level of 11 different APIs in the effluent water is about 36.96 milligrams per liter. These levels are a small percentage of the water outfall from the common effluent treatment plant (CETP). He suggests that the cumulative "fluoroquinolone (ciprofloxacin is a fluoroquinolone) concentration in ecotoxicological context is higher than the maximal therapeutic human plasma levels." Thus, the API concentrations at these levels have a toxicological influence on the environment, are a cause of concern, and should be controlled. We need to recognize that there may be very few or no individual and/or collective ecotoxicological standards for many of the chemicals produced. Discussion of their levels and influence is outside the realm of this document. Dr. Larsson suggests that the samples from PETL contained the "highest levels of pharmaceutical ingredients in any effluent." However, there is no identification and comparison to any "other" effluents.

Establishing Safe Levels

In our efforts to curb water and soil pollution, we will have to establish individual and respective safe toxic levels of various organics and control them to below defined levels in water bodies and soil around the producing plants. Recognition of the toxicity of these disease-curing chemicals presents us with two ways to reduce their levels below toxic levels:

1. Improve pharmaceutical processing and manufacturing technologies, which are considered inefficient and antiquated.[3]

2. Reduce organic levels in effluent water below their toxic levels.

The initial design capacity of the PETL wastewater treatment plant is not known. However, it is expected that necessary enhancements and provisions have been made to the plant over the years to keep the operating efficiencies at the optimum level. Table B.3 shows the operating results of the PETL wastewater treatment plant.

The Andhra Pradesh Pollution Control Board (APPCB) and other government bodies have to decide if the performance of the PETL operated CETP meets their set standards. From a pure numbers standpoint, it seems that there is reduction in the levels of BOD, COD, TDS, and TSS.

In addition, appropriate regulatory bodies have to decide on the toxicological impact of organics in the effluent water and the solids being sent to the landfill. Some of the background information that would be useful is not covered in the Larsson paper and is not available from APPCB4.

Background Information

1. No details about PETL are available.

2. Matrix Labs (Mylan Labs), Neuland Laboratories, and Aurobindo Pharma are fluoroquinolone producers sending their wastewater to Patancheru CETP. Along with other companies, they are shareholders of this facility.[5]

3. The current BOD, COD, TDS, and TSS standards Patancheru CETP has to achieve are not available. It is expected that they have to meet the standards set by APPCB.

4. There are no ecotoxicological standards for PETL CETP and many of the APIs.

5. There is considerable litigation about pollution due to Patancheru CETP.

Table B.3   Operating results of the PETL wastewater treatment plant.

	In mg/L	Out mg/L
BOD (biochemical oxygen demand)	1300	270
COD (chemical oxygen demand)	6000	1400
TDS (total dissolved solids)	9000	5000
TSS (total suspended solids)	500	300

Process Improvements Lead To Lower Waste

Since all APIs are a chemical and every drug is a formulation of chemicals to facilitate dispensing, implications of Dr. Larsson's work can be extended to any chemical produced anywhere. Individual organic component concentrations in the effluent point to the manufacturing efficiency of any API manufacturing facility worldwide and it needs a review.

Based on the concentration of organics in the effluent water Patancheru CETP outfall has about 45 kilos of ciprofloxacin per day. We do not know the concentration of the ciprofloxacin coming into the treatment facility. There are two assumed scenarios:

1. The wastewater treatment facility is not able to remove any ciprofloxacin from the incoming water.

2. Wastewater treatment is removing "X%" of the ciprofloxacin from the incoming water and the rest going with the sludge.

Scenario 1: The treatment facility is not able to remove any ciprofloxacin from the water and all of it is going in the effluent water. Based on 70% yield of the manufacturing process, one can calculate that the theoretical capacity of all the plants that manufacture ciprofloxacin to be about 150 kg/day and lose about 45 kg/day in the water. At $50 per kg, the dollar loss is estimated at about $0.8 million per year at active value and about $8 million per year at the drug counter value. If the plants can improve their yield from 70 to 85%, the increased revenue for the manufacturing plants would be about $0.4 million per year. This yield improvement would also reduce the organic loading of the water leaving their respective facilities and PETL.

Scenario 2: We do not know the ciprofloxacin concentration of the incoming water or the sludge leaving the site. If the wastewater treatment facility were able to remove, for example, about 50% of the incoming ciprofloxacin in sludge and the rest went with the effluent water, then the theoretical manufacturing capacity of the ciprofloxacin plants would be about 300 kg/day. Under these assumptions, plants lose 90 kg of ciprofloxacin per day at 70% yield. At $50 per kg., the API loss is estimated at about $1.6 million per year. Yield improvement of 15% would give additional revenue of about $0.8 million per year at the API level. Over the counter values are about 10 times these values. It is possible that many drugs and actives are pollutants in one form or other. We might be faced with a question of making a choice and balance between human needs, life extension, and environmental wellness. At some point, we humans will have to make a choice between feasible, acceptable, and affordable on each of these issues. Work similar to Dr. Larsson's could be done at other active pharmaceutical ingredient manufacturing, drug formulating, and chemical producing sites around the globe. It would be interesting to see the results of similar work and the potential recourse one would take to remedy the situation. The Patancheru study suggests that there is reduction in the value of parameters from the wastewater treatment. However, various lawsuits imply that the water and land around the facility is polluted. If the CETP operated by PETL is meeting the set standards then these lawsuits raise questions about the adequacy of the current water and soil environmental regulations for the area and around the world. If we have to meet ecotoxic standards then such standards have to be established for every chemical or their mixtures as each chemical has some ecological toxicity. As I indicated earlier, this would be an arduous and expensive task. I would not venture to speculate the expense or a finish date. If we have to reduce the organics below the toxic level then their levels have to be established. Remediation technologies will be needed to achieve the set limits. Some existing technologies could be used and others might have to be developed. We might need to change the water pollution standard for every water body in each country. This would have to be done for soil also. Implementation of various technologies might increase the selling price of our drugs multiple times and could make them expensive. Simple-to-use measurement technologies and instrumentation to track toxic levels that can be easily used by a trained technician will have to be developed. Most likely, we do not have such methods and procedures and it could be quite an expensive and time-consuming undertaking.

In the Meantime...

While the eco-toxic data is being developed, we have another opportunity to reduce the organic levels in the effluent. Costs related to this effort are lower and there are economic benefits. We need to improve process technology and manufacturing methods of API (chemical) and drug manufacturing plants. It is the obligation of everyone associated with the pharmaceutical and chemical industry, especially in process chemistry innovation, product and process development, commercialization, and manufacturing to improve the existing technologies, as they are not the best. Our pharma technologies are most inefficient and this has been echoed by the USFDA and others.[6-11] Improved manufacturing technologies have dual benefit of higher profit and reduced ecological damage while serving human needs. This might look difficult and challenging, but it is the least expensive option and has a quick pay back.

References

1. Joakim Larsson, D. G., de Pedro, Cecilia, and Paxeus, Nicklas. "Effluent from Drug Manufactures Contains Extremely High Levels of Pharmaceuticals." *Journal of Hazardous Materials*. Volume 148;Issue 3; September 30, 2007;751–755.
2. "Study Finds Significant Pollution from Indian Pharma Producers." *Chemical Week*. September 12, 2007;39.
3. Innovation and Continuous Improvement in Pharmaceutical Manufacturing. The PAT Team and Manufacturing Science Working Group Report. *http://www.fda.gov/cder/gmp/gmp2004/manufSciWP.pdf*.
4. APPCB List of Industries Region Wise. Accessed September 14, 2007. *http://www.appcb.org/list_of_industries_region_wise.html*.
5. Annual Reports of Matrix Labs; 2006/2007, Neuland Laboratories, and Aurobindo Pharma, 2006/2007. Accessed September 19, 2007.
6. Malhotra, Girish. "Batch or a Continuous Process: A Choice." *Pharmaceutical Processing*. March 2005;16.
7. Malhotra, Girish. "API Manufacture-Simplification and PAT." *Pharmaceutical Processing*. November 2005;24–27.
8. Malhotra, Girish. "Less Is More in API Process Development." *Pharmaceutical Manufacturing*. July/August 2005;50–51.
9. Malhotra, Girish. "QBD: Myth or Reality?" *Pharmaceutical Processing*. February 2007;10–16.
10. Malhotra, Girish. "Continuous Processes Maintain Profitability." *Drug Discovery and Development*. June 2007;30–31.
11. Malhotra, Girish. "Big Pharma: Who's Your Role Model, Toyota or Edsel?" *Pharmaceutical Manufacturing*. June 2007;40.

IMPLEMENTING QBD: A STEP-BY-STEP APPROACH

An In-depth Look at QBD and How It Can Help the Industry

Reprinted with permission from Pharmaceutical Processing (www.pharmpro.com)

We have to recognize that pharmaceuticals are nothing but chemicals that just happen to have a disease curing and life extension value. Quality by design (QBD) is an acronym we all believe in. It is an integral part of our daily life. We buy products using quality and price as the criterion. QBD is an essential part of business as it improves profitability. Almost every industry builds quality as part of their product design and manufacturing practice.

However, in the pharmaceutical industry quality has been achieved through "repeated inspections" as the product is produced. This has worked extremely well as we only see quality medicines on the shelves. The "repeated inspection" has an associated cost and we pay for these inspections as part of the overall cost of medicine. It would be helpful to review the roadblocks pharmaceutical companies have to overcome to implement QBD practices.

We need to recognize "what is a pharmaceutical process?" I consider the development of a drug for a disease to be a combination of 1) invention of a molecule, 2) testing of the molecule (active ingredient) for its efficacy, 3) production of the active pharmaceutical ingredient (API), and finally, 4) combination of API with excipients to produce a dose dispensable product, which could be a tablet, capsule, ointment, or a liquid product. Costs related to drug discovery and trials are necessary and have to be recuperated. They are not part of this discussion.

Since quality is not built into the pharma product design, "repeated inspections" are necessary. The FDA and other regulatory agencies acknowledge that the current manufacturing practices are antiquated. API manufacture falls short of the practices used in the chemical industry. We have to recognize that pharmaceuticals are nothing but chemicals that just happen to have a disease curing and life extension value. Techniques and formulation methods of APIs with excipients have been studied considerably, but they are still produced with "quality by analysis" methods.

Regulatory agencies would like pharmaceutical manufacturers to incorporate QBD in the manufacture of existing and new products. However, pharma companies face a dilemma and challenge to implement QBD practices as their products are highly regulated. Appropriate regulatory agencies have to be informed of any change in the current manufacturing process and/or quality. Depending on the change, requalification and re-approval might become necessary. This can be an expensive and time-consuming process.

Ethical drug producers have limited time in their patent life. QBD and re-regulatory approval might be un-economic for their existing products. Generic producers have the incentive to incorporate QBD, but depending on their end market may or may not see value in QBD. This is illustrated later.

Steps for QBD

Can pharmaceuticals be produced by having a QBD process? The answer is yes. It is necessary to incorporate the following in the development process. Steps might need alteration to suit individual needs of companies. Incorporation of the following might seem difficult at first, but once they are practiced everyday, they will become easy. QBD has to be a daily mantra and I have lived through such an experience. The long-term benefit is having the best and cost effective business process. It might even reduce regulatory complexities.

- Incorporate methods and thinking as if "every" drug molecule is going to be commercialized.
- Understand the chemistry and kinetics.
- Understand and implement physics, chemistry of the processes, their incorporation, and translation to an economic commercial process.
- Understand how to manipulate the chemistry and kinetics so that the commercial process is producing lowest cost but quality product with no rework needed.
- Understand and familiarize with the equipment and the technologies that are available and being used, not only in the pharmaceutical industry, but also in other industries, i.e., "cross-fertilize." This is important, as many times potent drugs do not require large batches.
- Make sure the "right" caliber expertise is available at your company to develop and implement the above improvements.

Since the collective benefits of the above are not totally understood, most of the time batch processes are an exact replica of the laboratory process. Lack of understanding of the above factors also results in investment higher than is necessary. For the same reasons, continuous processes are not considered for the manufacture of pharmaceuticals.

Reasons for Lack of QBD

The best way to understand the lack of QBD is by an illustration. Cost of API and excipients in the total cost of a drug are never discussed. It might be of value to understand these costs and their implications.

To every ethical and/or the generic pharma producer, the cost of API and excipients are a raw material cost. They are not concerned by the low yields of API or excipients. API and excipients are combined to produce a dose and sold at a price that the market will bear rather than what the market demands. The reason for such pricing is "our desire to extend life at any cost."

For the illustration, I have assumed the cost of an API that is under patent is $50.00 per kilo. It is also assumed that the combined cost of excipients is $25.00 per kilo. The API to excipient ratio in the tablet is one to nine. Based on these assumptions, the component cost of API and excipient for a 100-milligram tablet is 0.5 cents and 2.25 cents respectively.

It is assumed that the cost to formulate and package is 2 cents per tablet. This brings the total cost of a finished dose to 4.75 cents. If this tablet can be sold at $1.00 per tablet, the cost of API is only 0.5% of the selling price. Since this is a very low percentage of the selling price, there is no incentive for the ethical producer to change/improve their API manufacturing processes. Similarly, the cost of excipients is also a low percentage of the selling price.

Now let us assume that the patented drug becomes generic. The generic producer now sells the $1.00 per dose at $0.80 per dose. The price to the consumer has dropped but the cost of API on a percentage basis has gone from 0.5% to 0.625% of the selling price. Even though this is a 25% increase of the API component, it still is a very small portion of the selling price and is considered not worth worrying about.

Now let us consider what happens if Wal-Mart (or another similar seller) enters the fray. At $4.00 for a 30-day supply, per tablet price to the customer is 13.33 cents. The wholesale selling price of the generic single dose might be about 5–6 cents. Now the API and excipient raw materials are about 45–55% of the selling price.

As I have suggested earlier, the drugs under patent have a limited time before generics take over the market, thus ethical drug producers have limited incentive to lower costs of the API and excipients, as they are a minor percentage of their selling price. Generic producers have a higher benefit of QBD as they have more time and opportunity to lower costs.

How does one dictate QBD? Price control could be an answer. However, that would limit new drug discoveries and infringe on business freedom. Drugs sold in price control countries can drive QBD, but the lower costs may not be passed on to customers. Regulatory agencies can encourage the drug makers to implement QBD, but that is a slow process under the current laws. Sellers like Wal-Mart are the next closest thing to price controls. They can drive innovation. If innovation and productivity improvement are not a consideration for a company, then they might be thrust on them via "creative destruction," as suggested by Jacob Schumpeter of Harvard in 1942. Dr. Alan Greenspan defines this very eloquently.[1] "Creative destruction: A market economy will incessantly revitalize itself from within by scrapping old and failing businesses and re-reallocating resources to newer, more productive ones."

The best choice is for the drug makers to implement QBD themselves. It is necessary that the regulatory re-approval be simplified so that the manufacturers have an incentive to improve.

Since this is a significant percentage compared to 0.5% (for a patented drug) or 0.625% (for a generic drug), there should be incentive to reduce the raw material costs. Without players like Wal-Mart, there is no incentive for ethical or generic producers to improve their costs and implement QBD. API producers and excipient producers have an incentive to reduce their costs as they are competing with other producers based on their products being a chemical rather than a drug. The lack of the need to improve costs could be considered a "contrarian view" in a continuous improvement world.

Present manufacturing methods at drug companies are driven by the time constraint, as competition might have a similar product. There is no time to optimize the chemistry and engineering as applied to the manufacture of API development, excipients, and their formulation. Once the product starts to go through the regulatory channels, there is no time and change is difficult under the current regulatory environment.

Implement a lot of changes that will improve implementation of QBD process, reduce costs, and enhance product quality. They require an excellent understanding of chemistries and their translation to an economic process. A convincing argument about the improvements has to be made to the regulatory agencies and it can be made by show and tell. It may sound daunting, but having lived it, I can say it is easy after the first time. Due to escalating drug development and testing costs, pharma companies are looking at how they can reduce costs. We might see a temporary benefit to the pharma stockholders. However, the benefit of QBD applied to drug development and manufacturing is permanent and everyone will benefit.

Reference

1. Greenspan, Alan. *The Age of Turbulence*. The Penguin Press, New York, page 48.

PHARMA CONVERGENCE: CHALLENGES IN DRUG DEVELOPMENT AND MANUFACTURING METHODS

Published in CHEManager-Europe October 2008

A Changing Business

The global pharmaceutical world has been changing and its pace has accelerated in the last three years. Not only do they need to develop a new business strategy, they need to improve their manufacturing methods and technologies.

If the current scenario of "drying of the blockbuster pipeline" and generics increasing their market share holds, we could see most of the API manufacture, formulation, and clinical testing moving to low cost countries. Since laws of economics prevail, this could be considered inevitable. Ethical and generic companies have to develop and implement strategies that could give them the competitive edge and allow them to move forward on their chosen path. Since ethical and generic producers are adversaries, it would be interesting to see the playing out of the respective strategies. Let the match begin!

Ethical Pharmaceuticals

Major pharmaceuticals have developed and commercialized blockbuster drugs. However, they have not retained these drugs in their portfolio after the patents expire, as they have been busy developing new drugs. Producing patent expired drugs has not been part of their strategy.

Due to high profit margins, generics have taken over the patent expired drugs and have lately made every effort to do so through litigation. With aggressive entry of generic producers from Israel, Iceland, and India, turbulence in the pharma field has dramatically increased.

With the drying of blockbuster pipelines, escalating clinical trial costs and relentless pressure of generics to capture the market, ethical drug producers are trying to implement strategies to reduce their costs and retain their stronghold on the drug development chain. Some of the strategies being implemented include the following:

- Outsource drug development
- Outsource active pharma ingredients (API) manufacture and formulation
- Synergize small molecules and/or biotech combinations
- Acquire small/large biotech companies
- Whatever else works, i.e., collaborations.

Some of these strategies might work as a short term fix to retain profits, however, the long-term impact of these strategies is going to be significant. The biggest consequence is going to be the shift, disappearance, and/or reduction of the knowledge base from "major pharma" companies to the outsourced companies. Since the outsourced companies are in low cost countries, they have dual benefit of the above relationships. It makes them intellectually and financially stronger to become formidable generic competitors. We are beginning to see this happen.

Generic Pharmaceuticals

Generic pharmaceuticals are enjoying what I will call the best of all worlds. They are basking in unprecedented growth. I do not believe any of the financial analysts and pundits would have predicted this in the beginning of 2005. Customers would like to have drugs at lower prices, generics are able to fulfill this need in every market, and as a result, the demand for generic drugs has increased. This surge has increased generic business dramatically in recent years. They have utilized profits to grow organically and acquire sites that are being shed by API producers and formulators at significantly low costs. They have also benefited from the technology and intellectual property that comes with these acquisitions. Strategies being implemented by the generic companies are unconventional and this is causing additional turmoil in the pharma field.

The Future and Strategies

Pharmaceutical companies have achieved handsome profit margins by inventing new drugs and by producing generics. Customers have paid for every inefficiency in the development, clinical testing, manufacturing, and supply chain. Since pharma companies have been able to make respectable profits there was never a burning need to minimize the costs of each step. Everyone has been comfortable in their respective arenas. However, the drying of the blockbuster pipeline and generic companies trying to encroach on the playing field of ethical companies is changing the market dynamics.

The prices consumers pay for drugs in the U.S. and some other countries are not market driven but rather driven by what the market can bear. Many consider these prices high and are getting low cost drugs every way they can, e.g., from Canada, Mexico, imports, and/or the Internet. This has led to considerable debate and discussions as healthcare costs increase. Wal-Mart and a few other companies are offering drugs at low prices. This puts pressure on companies in the supply chain to continuously lower their costs. Therefore, companies will have to consider and implement new strategies.

If the major pharma companies are not able to develop new blockbuster or biotech drugs, they could start making generic drugs. This could lead to consolidation and formation of "mega" companies. My definition of a "mega" merger is a combination of an ethical and generic company to be players in both markets. These mega companies will not only develop new drugs, but will also have to make every effort to retain the patent expired drugs as part of their portfolio. If this happens, every step of the supply chain, especially manufacturing technologies, would be critically evaluated and methods implemented to reduce costs. The business model of mega companies could be a combination of market and consumer driven companies trying to maximize their market share. This should reduce global healthcare costs.

The India government has announced an innovative drug discovery program combining global IT firms (Sun Microsystems), researchers (Royal Society of UK, Imperial College of London, Medicine Sans Frontiers, etc.), companies, and young minds at India's scientific laboratories to invent drugs at a fraction of the cost of a multinational company (MNC). An open platform of drug research like Linux development is an interesting and innovative concept and path. Success here would genericize and commoditize pharmaceuticals and add additional pressures on pharma companies to implement technology improvements to reduce costs. Other business models will emerge. I expect that more than 50% of the pharmaceutical market will become a commodity market in the next five years and we will see prices drop.

Manufacturing Methods

Improvement of manufacturing technologies has not been part of any business model. In the last few years, there has been a considerable amount of discussion on the need to improve the manufacturing technologies, nevertheless, progress has been very slow.

New business models for ethical and generic pharmaceutical companies will have to include improvement of their manufacturing technologies. Today, active ingredients and formulation are dictated by "quality by analysis" methodology. It is not the way of the future. Pharmaceutical companies have to move to quality by design (QBD). QBD is being talked about in the pharma world but it needs to be put into practice. Specialty chemicals, petrochemicals, and other industries have produced products following QBD.

Technologies to achieve QBD exist, but need to be adopted. Improving manufacturing practices and QBD is not difficult. It requires discipline and dedication. Implementing QBD methods will change the landscape and it will be interesting to see what develops.

PHARMACEUTICAL COSTS, TECHNOLOGY INNOVATION, OPPORTUNITIES, AND REALITY

Reprinted with permission from Pharmaceutical Processing (www.pharmpro.com)

As long as companies are able to achieve the above objectives, any inefficiencies in product development, marketing, and manufacturing processes become irrelevant as the related costs are absorbed. Until recently, the priority has been to invent a new molecule, get it approved by regulatory authorities and get it to market. If profits are threatened, the development of additional and alternate markets is the first choice to regain revenue. With upcoming losses due to patent expirations, we are seeing the start of consolidation, acquisition, and relationships with biotech companies and with companies in the developing countries. Branded (ethical) pharmaceutical companies have believed in product innovation rather than process innovation. Based on pure financial numbers, this has been their ethos. Due to the need to develop new molecules and speed them to market, the pharmaceutical business model has had little incentive to develop innovative process development and manufacturing technologies. In addition, the current regulatory conditions are also not conducive to process innovation. Technocrats and bureaucrats would most likely say that should not be the case now or in the future.

Regulatory oversight for the pharmaceutical industry is critical and necessary. In recent years, regulatory agencies have promoted innovation in process development, API (active pharmaceutical ingredients) manufacturing, and formulation methods. Companies might try to innovate, but the financial justifications for innovation are weak unless the API and formulation costs are totally out of line. For innovation to occur and to hold, it has to come from within rather than being driven by regulatory bodies and others.

As good corporate citizens, pharmaceutical companies should have the best technologies so that their customers' needs are met, profits are maximized, and environmental impact is minimized. Most would agree that pharmaceutical companies do deliver on new drugs, service, and profits; however, their process development and manufacturing technologies are not state of the art. They are low on the priority chart in the pharmaceutical business. We have seen, and will continue to see, minimal effort in this area. The rationale for this posturing is discussed and explained later in this article.

The best way to discuss this is to review the pharmaceutical business model and the costs and various relationships from a different perspective. This discussion is focused on small molecules but could be extended to pharmaceuticals in general.

What Is a Drug?

Drugs are a combination of active pharmaceutical ingredients (APIs) and appropriate inert excipients. Pharmaceutical companies screen and explore molecules that will cure a disease and select the best ones for further testing. If a molecule cures a disease with minimal side effects, it is marketed. Although companies do manufacture and formulate the drugs, their manufacturing technologies are not current.

Cost of an API

To understand the pharmaceutical business model, it is worth understanding the price relationship between the API, the tablet, and the drug market price. Relationships between brand (ethical) and generic pharmaceutical companies also need to be considered. My overview is very general and extremely simplified. I will call my review "Everything we want to know about the cost of active pharmaceutical ingredients (APIs) and the corresponding tablets but are afraid to ask."

An API is an organic chemical that has a unique property: being toxic to specific bacteria in order to cure a disease. Since APIs are a fine/specialty chemical, like any other chemical, their factory costs can easily be estimated. Using a defined stoichiometry, any chemical engineer and/or chemist well versed in estimating factory costs and manufacturing methods can calculate the costs with reasonable accuracy. Similar methods can be used to estimate the factory cost of a tablet. If the factory costs are higher than the cost estimates, it suggests that there are opportunities to improve manufacturing technologies and thereby profits.

Factory cost of an API 2-benzhydrylsulfinyl acetamide (provigil/modafinil) is used in the illustration. Stoichiometry is outlined in U.S. patent 6,875,893. Although the yield and operating parameters of the process described in the patent are not optimized, the process envisioned is based on what I will call a "quality by design" process. Some of the processing steps that will improve productivity and throughput, based on my experience and knowledge, are included in the cost estimate.

A contract manufacturing organization (CMO) would sell the API to a formulator who in turn would sell it to a pharmaceutical company. Cost information in Table B.5. is used to illustrate the API, tablet, and average wholesale selling price (AWP) relationship.

Raw material prices were obtained from different suppliers. Conversion costs are based on my experience and what I consider to be reasonable for a specialty chemical product. Factory cost of the acetamide is about $8.34 per pound. It is sold at $13.91 per pound ($30.60 per kilo) with 40 percent profit margin to a formulator. This API is converted to tablets and sold at $210.65 per kilo. Tablets are bottled and sold by number rather than by weight. Table B.6. illustrates the cost component of the API in each tablet and the cost of bottled tablets.

All of the above costs are based on having a plant in the U.S. If the plant were located in India or China, due to currency parity and other factors, the costs would be lower by at least 15–20%.

Table B.4 Potential bulk selling prices of API and tablets.

2-Benzhydrylsulfinyl acetamide (Modafinil cost) USP 6,875,893					
Chemical	Moles	Mol Wt	Pounds	Price $/lb	$
API Cost					
Benzhydrol	1	184	83.72	3.18	266.39
Thiourea	1.2	76.1	41.55	1.00	41.55
HBr	1.2	81	44.23	0.80	35.38
KOH	2.88	56	73.38	0.55	40.36
Chloroacetamide	1.5	93.5	63.82	1.30	82.96
Acetic acid	3.2	60	87.36	0.25	21.84
H_2O_2	2	34	30.94	0.27	8.45
Methylene chloride		84	5.00	0.35	1.75
Crystallization solvent					2.00
					500.68
Yield, %	80.4				
Modafinil		273.35	100		
				Raw material cost $/lb	5.01
				Conversion cost $/lb	3.34
				Factory cost $/lb	8.34
	Margin	40%		API sell price $/lb	13.91
				Excipient cost $/lb	3.48
				Formulation expense $/lb	17.38
				Packaging $/lb	22.68
				Tablet factory cost $/lb	57.45
	Margin	40%		CMO tablet sell price $/lb	95.75

Table B.5 API cost per tablet and finished tablet cost.

Milligrams	200
Number of tablets/kilo of API (80% yield)	4000
API $/kilo	30.60
API cost per tablet, cents	0.76
Finished tablet cost, cents	5.27

Table B.6 API needed for billion dollar sale at different AWP and dosage.

	Kilo/$ Billion	
Dosage, mg	AWP $ 10.00	AWP $ 3.00
200	25,000	83,333
100	12,500	41,667
50	6250	20,833
10	1250	4167
1	125	417
0.5	6.3	208

Annual API Requirement for a Blockbuster Drug

Branded/ethical or generic pharmaceutical companies will sell these tablets at a price of their choosing. Generally a drug is considered a blockbuster if its annual revenue exceeds $1 billion per year. Average wholesale selling price (AWP) of the 200 milligram tablet modafinil is $10.581. Pharmaceutical companies sell the tablets to other companies for further sale to consumers. Recently, 200 milligram provigil/modafinil tablets sold for about $15.00–$16.00 per tablet at a national drug chain. Based on Cephalon's annual report,[2] modafinil has sales of about one billion dollars and would be considered a blockbuster drug.

Figure B.1 illustrates the amount of bulk API that is needed to produce sufficient tablets to have $1 billion in revenue for different drug dosages. For simplification, it is assumed that the API and formulation yields are 80%.

For 200 milligram tablets to reach $1 billion per year in revenue, about 25,000 kilos of API per year is needed. If the pharmaceutical company's dosage and selling prices (AWP) are different, the needed API volume for a blockbuster drug will change correspondingly. This is discussed later.

Figure B.2 illustrates the cost of API per tablet at different dosages for API ranging in cost from $10.00 kilo to $100/kilo. A formulation yield of 80% is assumed.

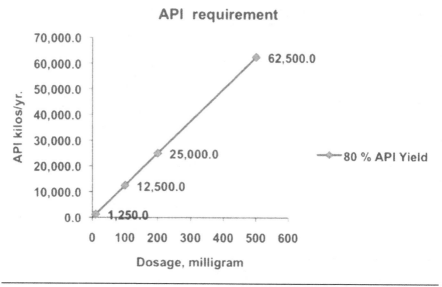

Figure B.1 API needed per year for different dosages.

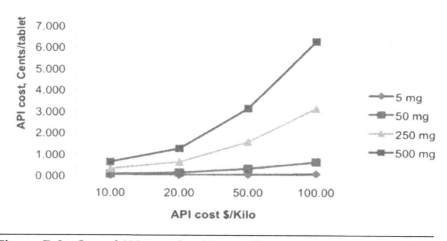

Figure B.2 Cost of API in each tablet at different dosages.

Based on the AWP, I would conclude that it is set at the highest level the market will bear, and pharmaceutical companies will recover and fund their costs related to marketing, drug trials, research of new molecules, etc. It has been said that with the low success rate of developing a useful drug, the current pricing methods are necessary. This might be true. However, we do need to review the relationship between AWP, the cost of API per tablet, and the cost of each tablet.

Based on the API manufacturing and tablet formulation cost estimate, the cost of API in each 200 milligram tablet is about 0.76 cents and the cost of each finished tablet is about 5.27 cents. Revenue generated by the sale of the bulk API and tablets would be about $0.77 million per year and about $5.3 million per year respectively, negligible compared to the yearly revenue generated by a blockbuster drug. Even if these API and formulation costs for each tablet were doubled or tripled, compared to AWP, they would still be minor.

With such a high monetary differential between AWP, API, and formulated drug costs, the impact of any savings due to manufacturing innovation will be inconsequential and will not have any after tax impact compared to after tax savings from improving marketing, clinical trials, and drug development methods. Thus, there is no incentive for branded or generic pharmaceutical companies to innovate API process development, manufacturing, or the tablet formulation technologies. Still, their technologies have to be the best they can be. Anything short will increase costs.

Branded (Ethical) and Generic Relationship

Generic companies have changed the playing field. Similar to branded companies, they also do not have any financial gain from technology innovation. This is explained as follows.

At present the generic companies do not have new molecule development costs, and their profit margins will be comparatively higher than the margins of brand companies. Due to lower expenses, generic drug AWP is lower than the brand pharmaceutical AWP, although they are still at the highest level. Generic AWP is also many multiples higher compared with API and formulation costs and can achieve their margins. Like branded companies, they do not have any incentive to innovate manufacturing technologies. ROI on any investment to improve technology, based on AWP, will not meet any financial norms, resulting in minimal or no investment.

Generic pharmaceuticals have found another creative method to increase their profits. This is through challenging the branded (ethical) patents.[3, 4] If a generic company challenges the ethical drug patent, the branded company has to defend it due to the revenue it could lose and subsequently prevent them from funding the discovery of new molecules. In a win-win negotiation, the ethical drug company offers money to the generic company for not entering the market until the patents expire, e.g., one time payment of $200 million[5, 6, 7, 8] (or whatever can be negotiated). The branded drug company may also make the challenging generic producer the contract manufacturer of the API and the drug tablet. This allows the generic company to make additional profits at the API and the formulation stages for "x" years until the patent expires, along with six months exclusivity. Even with these concessions, the branded company, based on the above pricing model, is able to generate sufficient monies for the development of new molecules, as these costs are amortized over "x" years. There is no significant effect on the cost of the tablet and AWP. There may be other relationships between the brand and generic companies that might not be public.

Once the patent expires, a generic company will produce and sell the drug after regulatory approvals at the highest price the market will bear. With these opportunities for the generic companies, we should not be surprised to see additional patent challenges in the coming years. Prescription generic drugs are priced lower than the branded drugs. Recently, this has started to change further as mass merchandisers like Wal-Mart or Target have started the selective selling of prescription drugs at 13.3 cents per tablet for one month's supply. Some of the drug and food stores have joined in with similar marketing strategies. This basically suggests that lower drug prices are feasible, and the marketers are profitable. If a generic company does not want to develop new molecules, in their efforts to expand their business they could significantly reduce their costs and selling prices through manufacturing technology innovation. If this happens, the game could change further. There are signs that this is slowly becoming reality.

Generic companies are in an enviable situation. As they gear up to form the Collaborate —> Cooperate —> Compete scenario, we will see drug prices stay at higher levels. Through Teva Pharmaceutical Industries, Ltd., we are seeing the application of this "3C" strategy.[9] Other companies are experimenting with this strategy and every success in this field is rapidly going to embolden other generics firms to participate.

Current Manufacturing Model

As stated earlier, the total cost of bulk API and bulk tablets is inconsequential compared to the revenue generated from the wholesale sale of the tablets. Even if the API costs are multiple times higher, branded companies have no interest to drive the costs down as their interest is inventing and marketing a new molecule.

If a pharmaceutical company were producing the API, any investment to improve manufacturing technologies would be for either of the following reasons:

1. Cost of a competitive API is lower than their molecule.

2. To appease the financial analysts, stakeholders, and/or regulators, the company is investing in upgrading its technologies even though there is no return on investment.

The production volume of 2-benzhydrylsulfinyl acetamide needed to generate $1 billion sales is about 25,000 kilo/year. This volume is not high enough to have the most innovative process. However, it is a business opportunity for companies producing the API. They can innovate manufacturing technologies, lower the API costs, and improve their profits.

Alternate Manufacturing Model

For manufacturing and technology innovation in pharmaceuticals, we need a disruptive paradigm shift. Since the small-molecule APIs are specialty chemicals, it would be productive to benchmark API costs, manufacturing methods, and technologies to respective specialty chemical methods. I would consider changing the current business model, where the small molecules are produced internally, to complete outsourcing of the API (specialty chemical) and formulation to companies who are "good to great." These companies have the financial incentive to improve their profit margins and will invest in technologies that will improve their productivity and deliver consistently high quality. Quality by design is their forte. They apply "Jugaad"[10] to minimize costs and to maximize quality and customer service to achieve their profits.

Independent of who produces the API or formulates the tablets, we also need to STOP repeated sampling and isolation at every intermediate step. Such a move will force companies to innovate and move from "quality by analysis" to "quality by design" manufacturing. In our endeavor, we cannot forgo good manufacturing practices. Based on my experiences, at times we are using an "armored tank" when a "bicycle" will suffice to produce API. We need to have complete command of the process rather than the process driving us. We have been, and are, victims of "analysis paralysis."

To get out of this situation, alternate manufacturing scenarios could be considered. Contract manufacturing organizations (CMO) can specialize in certain chemistries, unit processes, and operations and campaign their production. Based on the amount of API needed, some of the products could be campaigned in pilot plants to mini-plants. Modular plants could also be an excellent option for batch or continuous processes.

At 10 milligram tablet size, if the market grows from $1 to $10 billion per year, the API requirement will increase from about 1.25 metric tons to 12.5 metric tons per year. This will move production from a laboratory to a pilot plant. For a 200 or 500 milligram tablet drug, a mini-plant with a continuous process could be a distinct possibility. Better production scheduling may even lower the financial impact due to reduced in-process inventories resulting in improved cash flow. Laboratory scale can be at mini-plant scale so that the facility can also be modular and flexible enough to accommodate different unit processes and operations. Such methodologies due to the high price differential between the cost of API and AWP of the drug have never been considered, as there has been no need. Maybe its time has come.

Table B.3 illustrates the API requirements at two different AWP levels and different dosages for a blockbuster (a billion dollars per year) drug at 80% formulation yield. Specialty and fine chemical companies produce products at such levels and do a great job as they use the best manufacturing technologies available to produce products with high on-stream time and the highest product quality.

Innovation in manufacturing methods, improved process technologies, along with good manufacturing methods can deliver excellent processes and can handle low to high production rates. Better manufacturing technologies for API and formulation will improve profits at that level. For API, the improvements come in the way of improved yield, reduced waste, reduced and better solvent use, improved productivity, and improved business processes. For formulations they come with the blending, size uniformity and uniform admixing, etc. In the supply chain there are opportunities and one has to pick them. Improved technologies will also significantly reduce their carbon footprint. With fine/specialty chemicals practicing innovative processes, the question is why pharmaceutical companies haven't gone there? The answer is simple. There has been no need and the simplest thing we humans can say is "No. Can't be done," or "Prove it."[11]

Companies that are excellent in manufacturing API and formulating drugs should be the innovation driver rather than the companies who know how to invent and market only. Based on a purely financial basis, branded and generic pharmaceutical companies have the least incentive to innovate and excel in manufacturing technology innovation if they are producing APIs and formulating them to produce the drug. They might innovate due to regulatory/competitive pressures or if a "creative destructor"[12] challenges their turf. There is considerable discussion of process and technology improvement, but in the final analysis "return on investment" will drive innovation.

Irrespective of the sourcing options, the companies who sell the formulated drug have to be held accountable for the product quality that is ultimately consumed.

References

1. http://www.fdhc.state.fl.us/Medicaid/Prescribed_Drug/xls/drug_pricing_091005.xls. Accessed November 13, 2009.

2. Cephalon, Inc. SEC filings10-K February 23, 2009 and 10-Q October 28, 2009.

3. Hatch-Waxman Act. Accessed January 11, 2010.

4. Higgins, H. J. and Graham, S. J. H. "Balancing Innovation and Access: Patent Challenges Tip the Scales." *Science*. Vol. 326;370–371.

5. U.S. Trade Agency Wants to End Deals Delaying Generics. *www.Reuters.com*. Accessed November 14, 2009.

6. U.S. Bill a Challenge for Indian Generic Firms. *www.livemint.com*. Accessed January 12, 2010.

7. McDonald, K. D. and Mauk, J. E. Reverse Payments in Hatch-Waxman Cases and the Continuing Anti-Trust Patent Battle. Accessed November 10, 2009.

8. Federal Trade Commission Staff, Pay-for-Delay January 2010. Accessed January 13, 2010.

9. Acharya, Dr. Satish. The Productivity Tiger—Time and Cost Benefits of Clinical Drug Development in India, Life Sciences and Health Care Practice at Deloitte Consulting, Zurich. Accessed November 10, 2009.

10. Malhotra, Girish. *What is Jugaad (New Management Fad from India)?* December 10, 2009.

11. Martin, Robert L. and Riel, Jennifer. "Innovation's Accidental Enemies." *Business Week*. January 25, 2010;72.

12. Malhotra, Girish. "Pharmaceutical Manufacturing: Is It the Antithesis of Creative Destruction?" *Pharmaceutical Manufacturing*. 2008.

INDEX

Printed in the United States
By Bookmasters